CHEMISTRY TRIVIA

1. Our knowledge of theoretical ideas regarding matter originates with what nationality?

(a) Greeks (b) Romans (c) Chinese (c) Egyptians

2. Robert Wilhelm Bunsen was highly inventive. He developed the Bunsen burner and which of the following:

(a) spectroscope
(b) electrochemical cell
(c) grease-spot photometer
(d) absorptionmeter
(e) actinometer
(f) filter pump
(g) ice calorimeter
(e) effusion apparatus

3. What famous eighteenth century chemist met death on the guillotine?

4. It was frequently said that the greatest of all Sir Humphry Davy's discoveries was________________________.

5. Which country was the first to use natural gas for industrial purposes? (a) Japan (b) India (c) Persia (d) Russia (e) China

6. What scientist discovered the most elements?

7. If Gay-Lussac discovered Charles' Law, why is it named for Charles?

8. Moissan was awarded the Nobel Prize in 1906 for his discovery of certain synthetic gems. True or false?

9. The rod wax deposited on the pumps of oil wells led Robert A. Chesebrough to the discovery of what "miracle jelly"?

10. The earliest known university laboratory dates back to 1609 at the University of Marburg, at which time Johannes Hartmann opened a laboratory for students. What was its purpose?

11. Who was the first to change one element into another and when did this occur?

12. At what time did alchemy begin to be a formal discipline?

1. (a) Greeks.

2. Bunsen developed each of the items listed. (The actinometer was developed with his English student Henry Enfield Roscoe.)

3. *Antoine Lavoisier*.

4. *Michael Faraday* (One of the greatest English chemists and physicists, Faraday became Davy's assistant at the Royal Institution in London in March of 1813 and remained there for 54 years.)

5. The *Chinese* (about 1000 B.C.) (This was accomplished by obtaining natural gas from wells and then sending it through pipes made of bamboo.)

6. *Sir Humphry Davy*. (These elements are sodium, potassium, barium, calcium, magnesium, and strontium.)

7. Some years before Gay–Lussac's work, Jacques A. E. Charles had made the same observation. For this reason, he was given credit for the discovery.

8. True and false. Moissan was awarded the Nobel Prize in 1906 but not for his discovery of synthetic gems. The Prize was his for his preparation of pure fluorine and for development of the electric furnace.

9. From rod wax Chesebrough discovered *vaseline petroleum jelly*.

10. For *pharmaceutical preparations*.

11. It was *Rutherford* who in 1902 changed nitrogen to oxygen by sending gamma rays from radium through nitrogen.

12. Toward the fifth century of the Christian era.

13. A huge debt is owed to the Muslims for continuing chemical tradition after Greek culture deteriorated. How was this accomplished?

14. This Russian–born scoundrel, who was fluent in seven languages, entered into two bigamous marriages, one with Pepita Bobadilla (who later married Alfred Nobel). When his finances were low, he practiced chemistry. Can you name him?

15. The terms *plutonium*, *americium*, *curium*, *berkelium*, *californium*, *mendelevium* and *niobium* are sweet to the ears of Glenn T. Seaborg. Why is this?

16. To Chemistry's Nobel Prize winner of 1932 goes the credit for the artificial snow we enjoy on the ski slopes today. Can you name him?

17. Distillation became important early in alchemical history. Besides stills, several other apparatus were in use at that time. Can you name them?

18. What were the original seven known metals?

19. To date, what country has garnered the most Nobel awards in chemistry?

20. By what name was the Persian Al–Razi known? What did he claim to have prepared?

21. What is the *Rosenhutte*? The *Moor's Head*? What distillate did these inventions make it possible to prepare?

22. Would you believe the first known chemist was a woman?

23. The Sumerians, Babylonians and Egyptians used aromatics for what purpose?

24. The combined work of Perkin, Kipping and Lapworth has left a lasting impression on experimental and theoretical organic chemistry. There is also an interesting item in the private lives of the three. Any idea of what this might be?

25. Avicenna is still considered the "prince of all learning " by Muslims today. What was Avicenna's great contribution to modern–day science?

13. As conquerors, the Muslims nurtured and adopted the chemical knowledge of the Persians, Egyptians, Syrians, Greeks and other nations who had been overrun by Muslim armies after the death of Mohammed.

14. *Sidney Reilly*, Ace of Spies.

15. Seaborg had a hand in discovering each of these elements.

16. We can thank *Irving Langmuir* for the snow nature forgets.

17. Pans, beakers, flasks, mortars and pestles, funnels, sieves, tongs and crucibles.

18. Copper, lead, tin, iron, gold, silver and mercury.

19. As of 1985, the U.S. just barely edged out Germany; Britain ran a close third.

20. The Persian Al–Razi was known as *Rhazes* in Medieval Europe. It was he who claimed to have prepared the Philosopher's Stone.

21. The *Rosenhutte* was a conical alembic made of metal; the *Moor's Head* was a glass or pottery vessel so designed that water might be placed around the alembic. These inventions made possible preparation of alcohol distillates of high concentrations.

22. It's true. Over 3000 years ago, women were in charge of making perfume, the composition of which constitutes a chemical reaction. The earliest mentioned perfumeress was Tapputi, mistress of the Belatekallim household in ancient Mesopotamia. Some of her successors were Mary the Jewess, Paphnutia the Virgin, and Cleopatra the Alchemist.

23. Aromatics were used to treat the sick.

24. Perkin, Kipping and Lapworth married three sisters.

25. Avicenna planned inductive and experimental approaches to determine those conditions which produce observable results.

26. What is another name for *aqua ardens*?

27. What are the original names for the following formulas?

H_2SO_4 HNO_3 $HNO_3 + HCl$

28. Who originated the heated water bath as an effective method for gentle operations?

29. What is the scholarly name for the Swiss physician Theophrastus Bombast von Hohenheim?

30. What is the name of the individual responsible for the systemization of Paracelsian medicine into iatrochemistry?

31. Georgius Agricola, a physician of Saxony, lived from 1494–1555. Why is he remembered today?

32. Can you identify the following alchemy symbols?

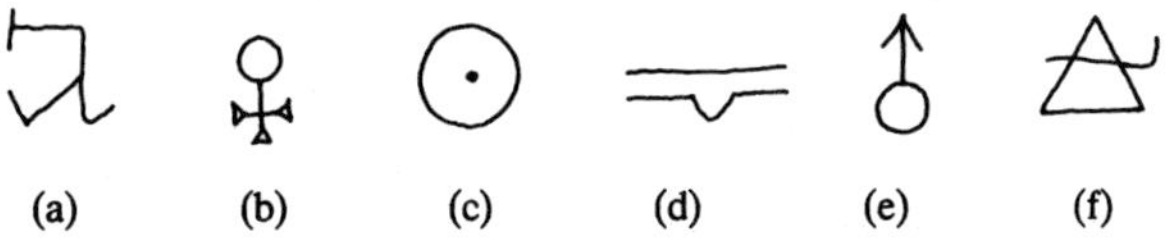

(a) (b) (c) (d) (e) (f)

33. The book *Beschreibung Allerfurnemisten Mineralischen Ertzt und Bergkwerksarten* was published in Prague in 1574. By whom was it written and what was its subject?

34. In what century were arsenic, antimony, bismuth and zinc recognized as metals distinct from the seven already known?

35. Christoph Schurer of Saxony introduced the use of what compound in the production of blue glass in 1540?

36. Who pioneered scientific thinking away from the Aristotelian syllogism?

37. In 1637 a Frenchman introduced the principle of systematic doubt. What was his name?

38. Angelo Sala considered fermentation as a regrouping of elementary particles to form new substances. True or false?

26. *Aqua vitae.*

27. *Oil of vitriol, aqua fortis* and *aqua regia*, respectively.

28. *Mary the Jewess.*

29. *Paracelsus* (Philippus Aureolus).

30. *Franciscus Sylvius* (Franz de la Boe).

31. Agricola was the first great mineralogist and one of the most important men in the early history of geology.

32. (a) borax, (b) copper, (c) gold, (d) to precipitate, (e) iron, (f) air

33. The book was written by *Lazarus Ercker*. The subject was mining and smelting problems with emphasis to assay methods.

34. Seventeenth century.

35. *Cobalt.*

36. *Francis Bacon.*

37. *Rene Descartes* (in *Discours de la Methode*).

38. True.

39. WHO WAS IT???

(Select answers from below)

a. Who first discovered uric acid in the excrement of birds?

b. A professor of geology at the Ecole des Mines conceived the idea of plotting the elements according to atomic weights on the surface of a cylinder. Who was he?

c. This Scotsman began the serious study of chemistry in 1856. Two years later, he published the symbol of the valence line to illustrate the linking of atoms with each other. After that, nothing further was published by him. Who was he?

d. He studied under Kekule and Baeyer. In 1875 he discovered phenylhydrazine, a valuable reagent for aldehydes and ketones. Although his fourth son became important in organic chemistry, the loss of three sons during World War I caused him to commit suicide. Who was he?

e. He was born into slavery. When a baby he and his mother were stolen one night by a band of raiders from their quarters. His master bought him back but his mother was never located. This remarkable man never looked for fame, but great men from around the world sought his advice. Though not a chemist by profession, he won the Spingarm medal in 1923 for distinguished service in agricultural chemistry. Who was he?

f. The law of simple multiple proportions asserts that when two elements combine to form more than one compound, the weights of the combining elements are in the ratios of small integers. Who established this law?

g. Who was it that showed methylethyl ether could also be produced from ethyl iodide and potassium methylate?

Selection of Answers:

A. E. Beguyer de Chancourtois
Emil Fischer
John Dalton
Alexander William Williamson
Louis Nicolas Vauquelin
George Washington Carver
Archibald Scott Couper

Correct Answers to 39.

a. Vauquelin.

b. Chancourtois.

c. Couper.

d. Fischer.

e. Carver

f. Dalton.

g. Williamson.

40. Johann Baptista van Helmont made progress toward understanding chemical entities. He

(a) believed water must be the basis of all things.
(b) showed that gold could be recovered from acids in which it was dissolved.
(c) recognized that a unique gas might be produced in several ways from different sources.

41. For what theory is the Englishman Robert Boyle known?

42. Two men are given credit for having laid the foundation of qualitative analysis. What are their names?

43. In *Physicae Subterraneae* (1667) ______________________ concluded that bodies are composed of three earths – terra lapidea (virtreous), terra mercurialis (mercurial), and terra pinquis (fatty).

(a) Georg Ernst Stahl
(b) Johann Joachim Becher
(c) John Mayow

44. The English biologist Stephen Hales was interested in plants. What was his major contribution to science?

45. It was a Russian physicist who recognized that heat was due to molecular motion, considered incorrect Boyle's idea of fire particles in combustion, denied the existence of phlogiston, and believed that matter could neither be created nor destroyed. All of these theories would be advanced by Lavoisier some forty years later. Can you name this eighteenth century Russian physicist?

46. Joseph Black studied at the universities of Glasgow and Edinburgh. Where was he later professor of chemistry?

47. Hydrogen was discovered in 1766 by what Englishman?

48. Joseph Priestley had no formal training as a scientist. True or false?

49. Priestly shares credit for oxygen's discovery with what Swede?

50. What *is* the Phlogiston theory?

40. (a) water.

41. *Boyle's Law* (which states that the volume of a gas at constant temperature varies inversely with its pressure).

42. *Otto Tachenius* and *Robert Boyle*.

43. (b) *Johann Joachim Becher*.

44. Hales observed that plants absorb air through their leaves. He conducted many experiments involving the collection and measurement of fixed air in solid substances by means of the pneumatic trough. In other studies, he measured the water consumption of plants, the pressure of sap rising in stems, the blood pressure of animals, and the speed at which blood circulates in the body.

45. *Mikhail Vasil'evich Lomonosov*.

46. Both. (Joseph Black succeeded William Cullen at the University of Glasgow in 1756; in 1766 he succeeded him at Edinburgh. There Black spent the remainder of his life.)

47. Hydrogen was discovered by *Henry Cavendish*.

48. True.

49. Priestly shares credit with *Carl Wilhelm Scheele*.

50. The Phlogiston theory sets forth that a substance called phlogiston escapes when a material burns, or when iron becomes rusty. (This theory was advanced by Georg Stahl, a Prussian chemist and physician.)

51. The term "oxygen" derives from the Greek. What does the term mean?

52. Who was it that sought to create a system in which water was the basis of all things?

53. On opposite sides of the globe, in August of 1779, Jons Jakob Berzelius and Benjamin Silliman were born. Can you give their respective countries of birth?

54. The teachings of M. H. Klaproth were of great importance in laying down the principles of sound analytical procedures. Klaproth either discovered or verified the discovery of zirconium, uranium, tellurium and titanium. What is his Christian name, country of birth, and in which university did he achieve prominence?

55. When and under what circumstances was zirconium metal first isolated?

56. Uranium was named for the planet Uranus (by Klaproth). What was tellurium named for and by whom?

57. Louis Nicolas Vauquelin was professor of chemistry at the College de France, at the Jardin des Plantes, and at the Ecole de Medicine. Do you know what he discovered?

58. William Hyde Wollaston was the son of an English clergyman. At the age of thirty–four he abandoned his chosen field to devote his life to science. In what field did Wollaston first become established? Which elements did he discover?

59. Osmium was named because of its odor; iridium because of the colors of its salts. Do you know who discovered these two elements?

60. Ruthenium was the last discovered of the platinum group metals. True or false?

61. What was the name given Antoine Lavoisier's first modern textbook of chemistry?

62. Claude–Louis Berthollet belongs to the great chemists of all times. He concluded that affinity alone cannot determine the direction of a chemical reaction. These conclusions lead to the law of __________, which was not formulated until well after Berthollet's death.

51. Oxygen means *acid former*, that is, to form or beget acid.

52. *Thales of Miletus*.

53. *Berzelius* was born August 20, 1779 in (Walversunda) Sweden; *Silliman*, August 8, 1779 in (Trumbull, Connecticut) U.S.A.

54. *Martin Heinrich* Klaproth was born the same year as Lavoisier in Wernigerode, Germany. He was the first professor of chemistry at the University of Berlin.

55. Zirconium metal was first isolated in *1824* when Berzelius heated *potassium zirconium fluoride with potassium*.

56. Tellurium was named for *tellus*, the Latin word for earth, by its discoverer, *Franz Joseph Muller von Reichenstein* of Austria in 1782.

57. Vauquelin discovered *chromium* and recognized the oxide of *beryllium*.

58. Wollaston enjoyed a prosperous *medical practice* before turning to science. He discovered the elements *palladium* and *rhodium* (1803). [Though palladium was named after Pallas, an asteroid, the name Palladium is the name of a statue of Pallas Athena. (Greek mythology tells that it was sent from Heaven by Zeus to Ilos, the founder of Troy. As long as the statue was safely in Troy, the city was safe. During the Trojan War, the Greek hero Ulysses and his companion Diomedes stole the statue and Troy fell.) Rhodium has no such illustrious legacy; its name was derived from the rose color of its salt.]

59. *Smithson Tenant* (In 1803 Tennant noted that a black residue remained after crude platinum had been dissolved in aqua regia. He believed this residue to be metallic, and in 1804 demonstrated that it was in fact two new elements, osmium and iridium.)

60. True. (Ruthenium, the sixth of the platinum metals, was discovered in 1844 by Karl Karlovich Klaus at the University of Kazan, in the Ruthenium region of Russia. The name ruthenium was derived therefrom.)

61. *Elements of Chemistry* (published in 1789).

62. Law of *mass action* (formulated by *Marcelin Berthelot*).

63. What was Joseph Louis Proust's most significant achievement?

64. From whence comes the word *ions* and by whom was the name given to chemical molecules?

65. The founder of the chemical atomic theory of the constitution of matter was born in September 1766 in the village of Eaglesfield to a poor hand–loom weaver. Throughout his life this most famous of Englishmen lived in relatively humble circumstances. Can you name him?

66. Both Joseph Louis Proust and Michel–Eugene Chevreul were born in the same city in France. Can you name this city?

67. During the first half of the nineteenth century, chemistry was dominated by the figure of Jons Jakob Berzelius. Not only was he one of the first to recognize the significance of Dalton's atomic hypothesis, but it was Berzelius who undertook the tedious analytical work which ultimately won acceptance of Dalton's theory. Where did he study and in what field did he receive his doctorate?

68. Though nearly all the chemical symbols suggested by Berzelius are in use today, a few have been changed. Can you name those which changed?

69. Four famous eighteenth century chemists were born in the month of December: Carl Scheele (December 19, 1742), Martin Klaproth (December 1, 1743), Benjamin Rush (December 24, 1745) and Humphry Davy (December 17, 1778). Can you cite their respective nationalities?

70. In the early part of the nineteenth century, the great center of French science was in a city just south of Paris. Can you name it?

71. In 1802 Charles' law, the law that deals with the effects of temperature on gases, was discovered. Who discovered Charles' law?

72. Boron was discovered in 1808 by:
(a) Louis Jacques Thenard of France.
(b) Joseph Louis Gay–Lussac of France.
(c) Sir Humphry Davy of England.

63. Proust's most significant achievement was the experimental proof of the law that chemical substances combine to form chemical compounds in constant proportions to weight.

64. *Ions*, the Greek word for wanderers, was the name given to the *migrating parts of chemical molecules* by the Englishman Michael Faraday.

65. *John Dalton* [Among other things, he proposed an atomic theory of matter (about 1803) and also investigated color blindness, which he had.]

66. Proust and Chevreul were both born in *Angers*, France.

67. Berzelius studied *medicine* at the *University of Upsala*. He received his doctorate of medicine in 1802. [In 1817 he discovered selenium (from *selene*, Greek for moon). It was Berzelius who suggested using the initial letter of the Latin name for each element because Latin was used more widely than any other language for scientific terms.]

68. Chromium (originally Ch) to Cr; Tungsten (Tn) to W; Columbium (Cl) to Nb; Platinum (Pl) to Pt; Palladium (Pa) to Pd; Manganese (Ma) to Mn; Beryllium (formerly called Glucinum, Gl) to Be; Magnesium (Ms) to Mg; Sodium (So) to Na; Potassium (Po) to K; and Chlorine (formerly called Muriatic radical, M) to Cl.

69. Scheele and Klaproth were Germans; Rush an American; Davy, English.

70. The great center was at *Arcueil*. [In 1805 Napoleon approved the formal organization of the Societe d'Arcueil. Its members included Berthollet, Laplace, Arago, Berard, Biot, Claude Berthollet's son (Amedee), Chaptal, Collet-Descostils, de Candolle, Dulong, Gay-Lussac, Humboldt, Malus, Poisson, and Thenard.]

71. *Joseph Louis Gay-Lussac* discovered *Charles' law*. (This law states that the ratio between the volume of a gas and its temperature remains constant if the pressure does not change, e.g., the expansion of a balloon with temperature rise.)

72. All three. *Thenard* and *Gay-Lussac* independently isolated boron at the same time as *Davy* (in England).

73. WHO DID WHAT???
(Select answers from below)

a. Henry Le Chatelier succeeded Moissan as professor of chemistry at the Sorbonne. For what discovery is he most remembered today?

b. In 1838 Louis Jacques Daguerre is said to have exclaimed: "I have seized the light! I have arrested his flight! The sun itself in future shall draw my pictures!" What had he discovered?

c. Why do we remember France's Rene Just Hauy?

d. Why is Otto Wallach remembered?

e. Herman Frasch performed his greatest experiment on an island in the middle of a mosquito-infested Louisiana swamp. What was it?

f. Albrecht Kossel studied medicine. His research in the field of protein chemistry led to the Nobel Prize in medicine in 1910. For what work was the Prize awarded?

g. For what reason do we remember Wilder Dwight Bancroft?

Selection of Answers:

(1) Oxyacetylene torches, used today in welding.

(2) For more than forty years a member of the staff of Cornell University, he sought to develop physical chemistry; perhaps his greatest service to chemistry was his founding of the Journal of Physical Chemistry. *Furthermore, he shed powerful light on many problems in colloid chemistry.*

(3) The pumping of underground sulfur to the surface in molten form.

(4) Cell chemistry, proteins and nucleic substances.

(5) Crystallography.

(6) Photography.

(7) Terpene chemistry.

Correct Answers to 73.

(a) (1)

(b) (6)

(c) (5)

(d) (7)

(e) (3)

(f) (4)

(g) (2)

74. Early chemists never failed to say what taste their newly isolated elements or compounds had. When in 1815 Gay–Lussac isolated the radical of prussic acid (since called cyanogen) did he depart from this habit?

75. In 1811, the Italian Amedeo Avogadro proposed that when equal volumes of different gases and vapors are put under the same pressure at the same temperature, they have the same number of molecules. This is known as Avogadro's law. True or false?

76. Anaximandros of Miletos conceived primary substance to be *apeiron*. What does this term constitute?

77. The relationship between specific heat and atomic weight (known as the law of Petit and Dulong) was reported by its discoverers to the Swiss Academy in the Spring of 1819. True or false?

78. The law of isomorphism states that compounds which crystallize in the same form are similar in chemical composition. Do you know who formulated this law?

79. The first edition of the well–known *Handbuch der Chemie* was published in 1817–1819. Who authored it?

80. Frederich Stromeyer discovered ______________ in 1817.

81. Bromine was isolated in the mid–1820's by several chemists. Who was the first to isolate it and how was this accomplished?

82. Only four organic acids were known before pharmacist Carl Wilhelm Scheele's studies (c. 1777). What were they?

83. Leopold Gmelin believed that organic compounds required a plant or animal for their production. He further considered simple inorganic compounds to be composed of two elements; organic compounds to contain three. As a result, he misjudged three organic compounds as inorganic. Do you know which three they were?

84. This German chemist first obtained a medical degree. He contributed greatly to both organic and inorganic chemistry. One of his greatest achievements was the synthesis of urea. Can you name him and tell how urea was synthesized?

74. Had Gay-Lussac placed a single drop of prussic acid on his tongue, he would have immediately collapsed as if struck by lightning.

75. True. (Though for many years scientists doubted the validity of Avogadro's law.)

76. *Apeiron* means potentially everything but really is undetermined and undefinable. [Incidentally, Anaximandros' younger compatriot, Anaximenes, saw air as the primary substance; Heracleitos of Ephesos (beginning fifth century) determined fire to be primary.]

77. False. The law of Petit and Dulong was reported to the *French* academy.

78. *Eilhardt Mitscherlich.*

79. *Leopold Gmelin.*

80. *Cadmium.*

81. *Carl Lowig*, *Justus von Liebig* and *Antoine-Jerome Balard* each isolated bromine in the mid-1820's. It was Balard, however, who first isolated it from a deposit of salts from which sodium chloride had been separated.

82. *Acetic, formic, benzoic* and *succinic*. (The first two were produced by the distillation of vinegar and ants; the last two sublimed from gum benzoin and amber.)

83. Gmelin treated *methane, ethylene* and *cyanogen* as inorganic compounds. (Incidentally, Gmelin was the son of Johann Friedrich Gmelin, a pharmacist and chemical historian, and the brother of Christian Gottlob Gmelin, professor of chemistry and pharmacy at Tubingen.)

84. Urea was first synthesized by *Friedrich Wohler* when he attempted to synthesize ammonium cyanate. Urea was discovered when he isolated the peculiar white crystallized substance which had been obtained in the reaction of cyanogen and aqueous ammonia. It was then he knew urea could be made without the use of kidneys.

85. Joseph Louis Proust studied the saccharine juices of plants and identified three sugars they contain. What are they?

86. The presence of what acid in opium became the basis for the ferric chloride color test to detect opium poisoning?

87. What is Faraday's Law?

88. To what school were atoms accepted as a fundamental part of material philosophy?

89. Because of his investigations with fatty substances, soapmaking and candlemaking became practical operations. Can you name this French chemist?

90. Organic analysis was still considered highly alchemical at the beginning of the nineteenth century. True or false?

91. From 1826–1850, Hessian University of Giessen, Germany, became the foremost institution for chemical instruction in the world. Do you know who was responsible for this phenomenon and how it came about?

92. Whose words are these? "The age of gold which was the age of peace and perfection; the age of silver, less pure and less noble; the age of bronze; the fourth age which seems to refer to the Minoan revival the glorious remembrance of which had inspired Homer; finally the age of iron, the present age of sorrow, hatred, and strife."

93. The term *critical temperature* means that temperature above which a substance cannot be liquified by pressure. Do you know who coined this term?

94. What are the five regular solids referred to as the Platonic bodies?

95. The manufacture of gas dates back to the early 1600's. An alchemist is given credit for producing the first manufactured gas when he tried to make gold. What is his name and where did this event occur?

85. *Glucose, fructose* and *sucrose.*

86. *Meconic acid* (discovered by Fredrich Wilhelm Serturner in 1805).

87. *Faraday's law* (which gave the first clue to the existence of electrons) states that a mathematical relationship exists between electricity and the valence (combining power) of a chemical element.

88. The school of *Epicurus of Samos.*

89. *Michel Eugene Chevreul.*

90. True.

91. At the age of twenty–two, Justus von Liebig was placed in charge of the small Hessian University. He accomplished the facility's rise by his practical approach to chemistry, which attracted students not only from Germany but from France, England, Italy, Mexico and the United States.

92. There are the words of *Hesiod*, who lived after Homer, perhaps toward the end of the eighth century, B.C. The quotation is from his poem, *Work and Days*, which tells the story of the five ages of the world.

93. *Herman Kopp.*

94. Tetrahedron, cube, octahedron, dodecahedron, and eicosahedron

95. *Jan Baptista van Helmont* is given credit for the first manufacture of gas, which occurred in Brussels, Belgium.

96. Chloroform, a heavy, colorless liquid, has a sharp odor and a sweet taste. It was independently discovered in 1831 by three chemists: Eugene Soubeiran of France, Justus von Liebig of Germany, and Samuel Guthrie of the United States. What substances were involved in the synthesis of chloroform?

97. Liebig and Wohler proposed that formulas for compounds be listed as below. What are the correct formulas for each?

(a) benzoyl hydride $C_{14}H_{10}O_2H_2$
(b) benzoic acid $C_{14}H_{10}O_2(OH)_2$
(c) benzoyl chloride $C_{14}H_{10}O_2Cl_2$
(d) benzoyl cyanide $C_{14}H_{10}O_2C_2N_2$
(e) benzamide $C_{14}H_{10}O_2N_2H_4$

98. Alchemists used three symbols to represent a crucible. Which of the following are they:

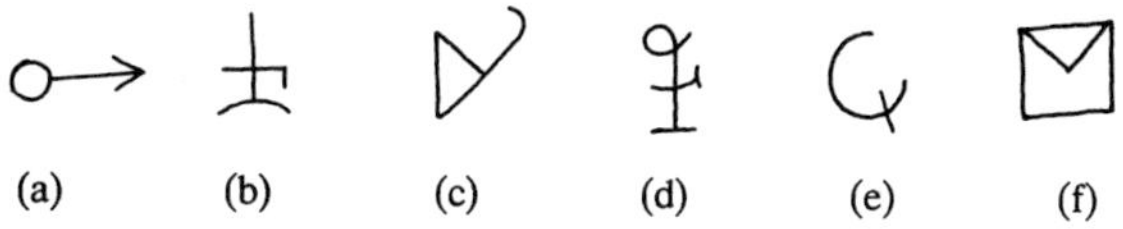

(a) (b) (c) (d) (e) (f)

99. The oldest still known to man (48 cm. high, 53 cm. in largest diameter), which was probably used to make perfumes, dates back to about 3600 B.C. Do you know where it was found?

100. It was a French chemist who invented an accurate means for determining nitrogen in organic substances. The method developed, which bears his name, remains in use today. Can you name this man?

101. In 1839 Charles Frederic Gerhardt proposed his theory of residues. This theory is also known by another name. What is it?

102. Why is Adolf Wilhelm Hermann Kolbe remembered today?

103. According to his own recollection, he was born the seventeenth child in Tobolsk, Siberia, on January 27, 1834. Among scientists, his name is almost a household word. Who is he?

96. *Alcohol* was treated with *bleaching powder.*

97. (a) benzoyl hydride C_7H_5OH
(b) benzoic acid C_7H_5OOH
(c) benzoyl chloride C_7H_5OCl
(d) benzoyl cyanide C_7H_5OCN
(e) benzamide $C_7H_5ONH_2$

98. (b), (c) and (f).

99. At *Tape Gowra* in Mesopotamia.

100. *Jean–Baptiste Dumas.*

101. Gerhardt's theory is also known as the *second radical theory* because the residues often have the same formulas as the dualistic radicals.

102. Kolbe was not only a chemist but later became editor of the *Journal fur praktische Chemie*. It was his work on organic formulas which led to a valuable means of detecting relationships between compounds.

103. *Mendeleev.*

104. During his lifetime, one of the greatest promoters of chemistry discovered the tetravalence of the carbon atom and the ability of atoms to link with one another. Oddly, he was a student of architecture at Giessen where he was influenced by Liebig to turn to chemistry. Can you name him?

105. Born on the 13th of July, 1826, his name would become engraved alongside Galileo, Torricelli, Volta and Galvani in the Pantheon. Yet today, his fame remains the treasure of Italy and all mankind. Who is this man and why is he regarded as a universal genius?

106. What other theory did this Italian develop?

107. As the twentieth century is considered that of the electron, the nineteenth century is considered that of the ____________.

108. Aristotle believed all things could be understood as unities of form and matter and set forth four principles in his philosophy of nature. These principles are as important for the course of chemistry as the concept of the elements. Material is the first cause. Can you name the other three?

109. Benjamin Silliman's position in American science was due not so much to his contributions to knowledge as to what?

110. In 1802 a French refugee established near Wilmington, Delaware, a factory for the manufacture of gunpowder. Who was he and what was the company he created?

111. Though he is remembered for his many inventions, this Englishman (in collaboration with Kirchhoff) discovered two elements. Can you name him and his elements?

112. The first Nobel Prize was given in 1901. The first American awarded the Prize (in 1914) was Theodore William Richards. What was the basis for this award?

104. *Friedrich August Kekule.*

105. The world of science is forever indebted to *Stanislao Cannizzaro* for the atomic theory.

106. Cannizzaro is also remembered for his *doctrine of valency*, the credit for which is often given to others.

107. The *atom.*

108. The other three are *efficient, formal* and *final.* (He considered the formal, efficient and final causes as form; the material cause as matter.)

109. Silliman was extremely enthusiastic in making his countrymen aware of science and was responsible for the training of a great number of young men who would later make their mark in education, medicine and industry.

110. The French refugee was *Eleuthere Irenee du Pont de Nemours.* The entity he created was the *Du Pont Company.*

111. With Kirchhoff, *Bunsen* discovered cesium (1860) and rubidium (1861).

112. Richards was awarded the Nobel Prize for having correctly determined the atomic weights of many elements.

113. In his periodic table, Mendeleev reasoned that the space following calcium should be closely related to boron and gave the missing element the provisional name *eka* boron (*eka* being Sanskrit for "first"). Similarly, he predicted the properties of *eka aluminum* and *eka silicon*. How, when and by whom were these three missing elements discovered?

114. His greatest accomplishments were in medicine in spite of the fact that he never formally studied this field. Can you name this French chemist?

115. What development facilitated the use of hydrogen sulfide?

116. In 1871 a significant contribution in the field of silicate analysis was made by J. Lawrence Smith. By means of this process he was able to convert two alkali metals to their soluble chlorides, leaving the fusing elements in insoluble form. How was this accomplished?

117. During the latter part of the nineteenth century, William A. Ainsworth and Son and G. P. Keller established firms to manufacture balances in the Rocky Mountain area of Colorado, U.S.A. True or false?

118. Karl Friedrich Mohr was primarily a manufacturer of pharmaceuticals, but he had an unusual talent for improving apparatus and procedures. Can you name some of his major accomplishments?

119. Between 1864 and 1879, Norwegians Cato Maximilian Guldberg and Peter Waage studied the whole problem of equilibrium reactions. How were these two men related to each other?

120. Alchemists must have been adept indeed to make any sense out of the strange pictures and symbols they used. Considering the norm, those symbols used to designate day and night are not too farfetched. Do you know them?

121. Sir William Henry Perkin entered the Royal College of Science at the age of fifteen. When he was just seventeen years old, he conducted an experiment in which he treated toluidine and then aniline salt with bichromate of potash. The result obtained was not artificial quinine but a dark, dirty precipitate. From this substance, he isolated the first dyestuff to be produced commercially from coal tar. Can you name it?

113. *Aluminum* was the first of the three missing elements to be discovered (in 1875) by Paul Emile Lecog de Boisbaudran; it is made from Bauxite ore. *Boron* was discovered (in 1879) by the Swede Lars Fredrik Nilson; it is found in the mineral euxenite (Nilson named it *scandium* because it had been observed only in ores found in Scandinavia). *Silicon* was discovered (in 1866) by the German Clemens Winkler who named it after his fatherland (*germanium*); it was found in argyrodite.

114. *Louis Pasteur*. He discovered the secret of steroisomerism, the protective inoculation of animals against anthrax and of humans against rabies, and was one of the pioneers in the field of infectious human diseases (to name but a few of his accomplishments).

115. The development of a *generator* (by P. J. Kipp, an apothecary in Delft).

116. Smith fused sodium and potassium in rocks with calcium carbonate and ammonium chloride, thus converting the two alkali metals to their soluble chlorides.

117. True. (Because of the heavy mining underway at the time, both Ainsworth and Keller established their firms in the Rocky Mountain area.)

118. Mohr made improvements in the analytical balance. He designed the specific-gravity balance, a reflux condenser, sets of cork-borers, and volumetric glassware (such as flasks and pipettes).

119. The two were *brothers-in-law*.

120.

day night

121. Sir Perkin inadvertently produced *mauve* or *aniline purple*.

122. Who developed polarization and what is the polarimeter used for?

123. How was the saponification number developed?

124. Born February 26, 1866, he was thirteen years old when Thomas A. Edison invented the electric light. He grew up in Cleveland, Ohio, and entered Case School of Applied Science in 1884. He once said, "I'd rather work for myself for $3,000 a year than to work for someone else and make $10,000." In 1889 he set forth to seek his fortune in what he and his partners grandly named The Canton Chemical Company. Can you name him?

125. By combining the kinetic theory of gases with the determination of the cohesion in Laplace's theory of capillarity, the Dutch scientist Johannes Diderik van der Waals derived which theory?

126. What is considered to be van der Waals' second greatest discovery?

127. Biot, discovered that turpentine, sugar, and certain other organic compounds rotate the plane of polarized light not only in pure form but in solution as well. What realization did this lead him to?

128. What is the *phase rule* and who developed it?

129. Jacobus Henricus van't Hoff and Joseph Achille Le Bel observed that when a carbon atom is attached to four different atoms or atomic groups, the four substituents can be arranged in two different ways. What is the result?

130. Van't Hoff's theories on the arrangement of atoms in molecules led to the development of which branch of chemistry?

131. A French scientist discovered radioactivity in 1896 while working with uranium ore. Was it? (a) Marcellin Berthelot, (b) Antoine Henri Becquerel or (c) Marie Curie.

132. In 1911, the British physicist Ernest Rutherford proposed a new theory of atomic structure. What is this theory?

122. The French physicist and crystallographer *Jean Baptiste Biot* in 1840. Polarization is the act or process of affecting light or other radiation in such a way that the vibrations assume a definite form. A polarimeter is used to determine the amount of polarization of light.

123. In response to a need for a test to distinguish butter from margarine, the saponification technique was first developed (in 1879) by the German Koettstorfer. (A saponification number is a standard value used in hydrolysis.)

124. *Herbert Dow,* founder of the Dow Chemical Company.

125. From this combination, van der Waals derived the *kinetic theory of the fluid state.*

126. Van der Waals' second greatest discovery was his *equation of state.*

127. Biot's discovery led to the realization that the optical activity of these compounds is due to a property inherent in the compounds themselves rather than to their crystal form.

128. The *phase rule* was developed by the American physicist Josiah Willard Gibbs. It provides a general relationship among the degree of freedom of a system f, the number of phases p, and the number of components c. This relationship is always $f = c - p + 2$.

129. Van't Hoff and Le Bel observed that such an attachment results in the molecules being *mirror images* of each other.

130. *Stereochemistry.*

131. (b) *Becquerel.*

132. *Theory of atomic structure* (which states that the atom has a positively charged nucleus surrounded by negatively charged electrons)

133. Who wrote the books on physical sciences *De caelo, de generatione et corruptione* and *Meteorologica*?

134. What is the Friedel–Crafts reaction?

135. The major metal organic compounds of interest in synthesis are the *Grignard reagents*, discovered in 1899 by Philippe Antoine Barbier. Why is the group labeled *Grignard* rather than *Barbier*?

136. In 1787 an amateur mineralogist discovered a rock he named *ytterite* (now known as *gadolinite*, after Johan Gadolin, the foremost Finnish chemist of his day). In 1803, Klaproth isolated a somewhat similar earth which he called *gerre ochroite*, while at the same time Berzelius and Hisinger discovered this earth and labelled it *cerium*. By the time the troublesome chemistry surrounding this earth was at end, a great many earths had been discovered. Any idea of what they are?

137. Although the German Friedrich Konrad Beilstein was best known as an editor, he succeeded Mendeleev as professor of Russia's Imperial Technological Institute. True or false?

138. American chemistry was an immature science at the turn of the twentieth century. However, there were at this time five native–born Americans who led the development of their country's chemistry. All five sought at least part of their education in Europe. One received the Nobel Prize in 1914. One became Chief of the Department of Agriculture's Bureau of Chemistry. Four became presidents of the American Chemical Society. Can you name them?

133. *Aristotle.*

134. The Friedel–Crafts reaction occurs when an intermediary compound is used as a catalyst, such that the hydrogen on the aromatic hydrocarbon nucleus is removed and replaced by alkyl groups. This reaction was introduced by Charles Friedel and James Mason Crafts in 1877.

135. Francois Auguste Victor Grignard was a pupil of Barbier at the University of Lyons, at which time Barbier asked him to study the reaction of magnesium organic halides. Once the usefulness of the reaction had been established, Barbier insisted that its credit go to Grignard.

136. *The rare earths.* (Following the name ytterite, the name *yttria* was applied. Then came *terre ochroite* and *cerium.* Thereafter, the soluble cerium oxide was called *lanthana*; *ceria* used for the insoluble oxide. Meantime lanthana was independently discovered in *mosandrite*, a new Norwegian mineral. Separation of lanthana resulted in two earths, *lanthana* and *didymia.* In 1843 the original yttria was separated into three earths: *yttria, terbia* and *erbia.* In 1878 erbia was separated into two earths: *erbia* and *ytterbia.* A year later saw ytterbia separated into *ytterbia* and *scandia.* Next, erbia was separated into three factions: *erbia, holmia* and *thulia.* Then another earth (*samaria*, from samarskite) was separated from didymia. Separated from samarskite was yet another earth, *gadolinia.* Samaria was further separated into *samaria* and *europia.* Didymia was again separated, to reveal *praesodymia* and *neodymia.* Holmia was then separated into *holmia* and *dysprosia.* Another fractionation of ytterbia revealed *neoytterbia* and *lutecia.* These two earths were then separated into *aldeberanium* and *cassiopeum.* The rare earth picture was then almost complete.)

137. True. (Prior to this time, when a vacancy occurred in the chair of chemical technology at the *Imperial Academy of Sciences*, it was Beilstein rather than Mendeleev who was elected to fill it.)

138. *Stephen Moulton Babcock*, the only one of the five who did not serve as president of ACS, was professor of agricultural chemistry at the University of Wisconsin. *Harvey Washington Wiley* became Chief of the Chemical Division in the U. S. Department of Agriculture. (Please see continuation, page 32.)

139. It was Oudry who demonstrated the identity of caffeine (from coffee) and thein (from tea). True or false?

140. Isolation of the first amino acid occurred in 1810. What was it and who discovered it?

141. Henri Braconnot obtained glycine from gelatin. However, because he failed to detect an element, glycine was regarded as a new sugar. What was the element he failed to detect?

142. What is meant by *allotropy*, a term introduced by Jons Jakob Berzelius in 1841?

143. The world's largest chemical company still operates from its original quarters on the banks of the meandering Brandywine River. Can you name it?

144. Many aspects of physics border on chemistry. Physicist Albert Einstein discovered the theory of relativity and the quantum theory of light – both of which are useful to a chemist. These two theories were not only revolutionary, but seemingly self-contradictory as well. Why is this?

145. What is meant by the "van der Waals attraction"?

146. Leo Hendrick Baekeland was born in Belgium. As a youth, he experimented with silver salts. Once he had migrated to the United States, he established a company to manufacture and sell a product of his invention. Eastman Kodak became interested and offered to buy the process. Baekeland was resolved to ask $50,000 and accept not a penny less than $25,000. To his astonishment, Kodak offered him $1 million for his process. What was it Baekeland invented and sold to Eastman?

147. When he was born in 1833, his family was in such financial distress that his father fled his nativeland to avoid prison. But when he died in 1896 he was one of the richest men in the world. Each year the interest from this man's vast fortune is divided into five equal parts and conferred upon that person who has made the most important discovery or invention to his field. Who is this man, how did he achieve his wealth and fame, and what is the award bequeathed in his will?

(138. continued)
Ira Remsen became head of the chemistry department of Johns Hopkins University and later its second present. *Theodore William Richards*, professor at Harvard, received the Nobel Prize in 1914. *Edgar Fahs Smith* was Vice–Provost and Provost of the University of Pennsylvania for a total of 25 years.

139. Partly true. Oudry isolated thein from tea but it was Runge who isolated caffeine from coffee.

140. The first amino acid to be isolated was *cystine*, discovered by *William Hyde Wollaston* in urinary calculi in 1810.

141. Braconnot failed to detect *nitrogen*.

142. *Allotropy* refers to those cases where an element is found in different molecular forms.

143. The world's largest chemical company is *Du Pont*.

144. Einstein's theory of relativity and the quantum theory of light seem contradictory because the former is intimately linked to the theory that light consists of waves, while the latter states that it consists somehow of particles.

145. *Attractive forces between molecules* that can be attributed to, or computed from, the attractive forces of molecules whose charge distribution enables them to behave like minute bar magnets.

146. Baekeland sold to Eastman Kodak his invention of a *photographic paper (Velox)* that could be developed in artificial light.

147. The party in question is *Alfred Nobel*. He achieved fame and fortune in *explosives*. The *Nobel Prize* is his legacy to mankind.

148. This compound (one of the simplest alkaloids) was contained in the cup of hemlock Socrates drank in 399 B.C. It was first isolated in 1831 and later synthesized by Albert Ladenburg, professor of chemistry at the University of Kiel. In 1965 the Nobel Prize in Chemistry was awarded to Professor R. B. Woodward of Harvard for his synthesis of alkaloids, of which this compound is an example. Can you name it?

149. Who was it that found atoms untenable on mathematical grounds?

150. Marie Curie was the first woman to receive the Nobel Prize in Chemistry (1905). The Prize was shared by a woman in 1935 and a woman received it in 1964. Can you name these women and the reason why they were awarded the Prize?

151. Christian Friedrich Schonbein discovered that cotton can be converted to an explosive substance. What is this substance?

152. John Dalton proposed symbols for the elements. Some of these are quite easy to intepret, such as:

Others are not:

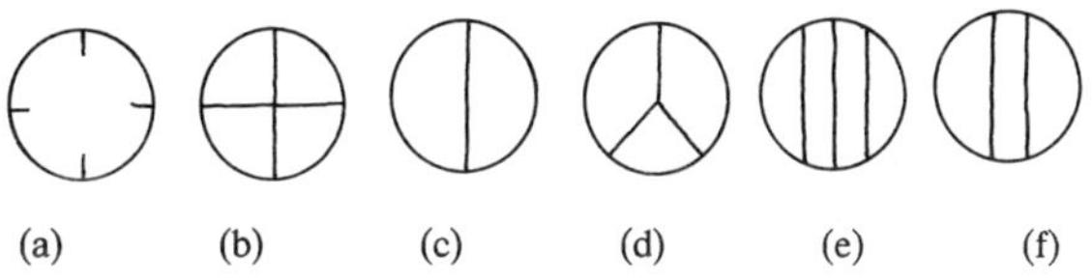

(a) (b) (c) (d) (e) (f)

Can you identify (a) through (f)?

153. Niobium was discovered in 1801 by the Englishman Charles Hatchett. From whence came the name *niobium*?

148. *Coniine*, which occurs in a member of the parsley family.

149. *Aristotle of Stagira*.

150. *Irene Joliot–Curie* (France) shared the 1935 Nobel Prize in Chemistry with her husband Frederic for synthesizing new radioactive elements. *Dorothy C. Hodgkin* (Britain) was awarded the Prize in 1964 for X–ray studies of compounds such as vitamin B12 and penicillin.

151. Schonbein found that *gun cotton* could be made by treating cotton with a mixture of nitric and sulfuric acids.

152. (a) strontium, (b) sulphur, (c) azote, (d) phosphorus, (e) potash, (f) soda.

153. *Niobium* was first considered identical to tantalum (named after the mythological Greek King Tantalus). Though the element was initially called *columbium*, in 1884 its name was changed to *niobium*, which was derived from Niobe, daughter of Tantalus.

154. In 1836, the Irish chemist Edmund Davy heated a mixture of calcined potassium tartrate with charcoal. He obtained a black mass that decomposed water with evolution of a gas. He had discovered acetylene. Yet it was Friedrich Wohler who, twenty-six years later, was actually credited with the discovery. Why was this?

155. Much of the work of the French chemist Ferdinand F. H. Moissan dealt with experiments using calcium carbide. His unsuccessful efforts to synthesize diamonds resulted in the design of ____________________.

156. Who is considered the father of botany?

157. Since the beginning of the Nobel Prizes in 1901, one man was nominated every year until his death in 1907 but each year passed over. Who was he and why was he passed over?

158. Chemist Lloyd Hall was a man forever blazing new pathways. He became the first black man to be chief chemist and research director for a major industrial company. For what accomplishments was Hall so widely regarded?

159. Count Rumford, Knight of the Order of the White Eagle, Lt.-General in the Service of His Most Serene Highness of the Elector-Palatine, the Reigning Duke of Bavaria, was born of poor parents in Woburn, Massachusetts. Later in life, he married the wealthy widow of a famous French scientist. What was Rumford's real name and who was the lady?

160. In response to a Russian nobleman's request to remarry, Czar Alexander II is known to have said: "Yes, Mendeleev has two wives, but I have only one Mendeleev." What circumstances surrounded this affair?

161. Ida Tacke Noddack is most famous as the discoverer of rhenium. She also played an important role in the story of nuclear fission. In what way?

162. On his gravestone appear the words, "He could have added fortune to fame, but caring for neither, he found happiness and honor in being helpful to the world". Can you name him?

154. Although Davy recorded his discovery of acetylene, he did not publish the results; because Wohler did publish a full account of his results, he is credited with the discovery.

155. *The electric furnace.*

156. *Theophrastos of Eresos.*

157. *Marcelin P. E. Berthelot* was passed over for the Nobel Prize because it was thought that younger men should be recognized to spur them to greater achievement.

158. Hall is remembered in the fields of biological and food chemistry for his work on meat curing products, seasonings, emulsions, baking products, antioxidants and protein hydrolyzates.

159. *Benjamin Thompson*, who married Lavoisier's widow.

160. According to the law of the Russian Orthodox Church at that time, a divorced man or woman could not remarry for seven years. Mendeleev divorced his first wife in 1876 and remarried before the official period was up. Although this made Mendeleev a bigamist, no action was taken against him because he was regarded so highly by the Czar.

161. Ida Noddack criticized Enrico Fermi's 1934 claim to having made the first transuranium element, stating: "When heavy nuclei are bombarded by neutrons, it would be reasonable to conceive that they break down into numerous large fragments which are isotopes of known elements but are not neighbors of the bombarded element." With these words, she conceived (before anyone else) the idea of nuclear fission.

162. This is the epitaph of *George Washington Carver*.

163. WHO WAS IT???
(Select answers from below)

a. Hundreds of persons yearly visit the remains of Karl Marx which lie in Highgate Cemetery, London. Nearby is the marked grave of the man who gave the world electrical power. Who lies there?

b. In the early eighteen hundreds, dentist Horace Wells' use of nitrous oxide in a demonstration of a tooth extraction to a group of students led to disaster and Wells' eventual suicide. This event, however, led to the use of ether as anesthesia. Who discovered the anesthetic properties of ether?

c. The father of American biochemistry was born in Vienna and raised in a small village in what is now the border of Yugoslavia.

d. The ideas which he submitted to the Swedish Academy of Sciences in 1885 (namely, the "laws of chemical equilibrium in the dilute, gaseous or dissolved state," "a general property of dilute matter," and "electrical conditions of chemical equilibrium") are now widely accepted.

e. He was educated at Cornell and at Heidelberg. Following twenty-eight years with the U. S. Geological Survey, he joined the National Bureau of Standards as Chief Chemist in 1908. In 1906 he was president of the American Chemical Society.

f. He lost his life in a fire during the operation of his process to fractionate light tar oil into benzene and toluene.

g. He solved the problem of the fourth valence of carbon by supposing an alternation of double and single bonds in the benzene ring.

Selection of Answers:

Michael Faraday | *William Francis Hillebrand*
Anton Alexander Benedetti–Pichler | *Jacobus Henricus van't Hoff*
Freidrich August Kekule von Stradonitz | *Michael Faraday*
Charles Blackford Mansfield

Correct Answers to 163.

a. Faraday

b. Faraday

c. Benedetti-Pichler

d. Van't Hoff

e. Hildebrand

f. Mansfield

g. Kekule

164. He is best known as a biologist, but he began his career in chemistry by placing molecules before a mirror to arrive at the conclusion that some are asymmetrical. Can you name him?

165. Ira Remsen was codiscoverer of saccharin. However, he is best remembered as an educator. Do you know why?

166. Stanislao Cannizzaro's determination of the atomic weights was fundamental to the next two important steps in chemical science. What are they?

167. Raymundus Lullus, often called King of the Alchemists, prophesized that gold could be made from baser less expensive metals. Today it is recognized that indeed gold can be artificially produced. How is this accomplished?

168. There are two tomb papyri dating from the end of the third century known as the Leyden Papyrus X and the Stockholm Papyrus. What do they represent?

169. Piccard is the name of a Swiss family of scientists. Two of the boys were twins: One became a physicist, the other an aeronautical engineer and chemist. Both were involved in balloon explorations. What were their respective Christian names and why is each remembered today?

170. The highest award bestowed by the American Chemical Society for accomplishment in chemistry is the Priestly Medal, named for Joseph Priestly. Priestly was born in England but spent his declining years in America. True or false?

171. The German chemist Victor Meyer's discovery of a way to synthesize aromatic acids with sodium formate was but the beginning of a long road to fame. Yet he took his own life while at the height of his career. Why?

172. It was the Dane, William Christopher Zeise, who first produced ethyl mercaptane in 1833. True or false?

164. *Louis Pasteur* (who said, "When we study material things, we soon recognize that they fall into two large classes....Those of one class, placed before a mirror, give images that are superposable on the originals; the images of the others are not superposable.")

165. It was Remsen who *founded graduate research chemistry in the United States.* (Constantine Fahlberg, who discovered the sweetening properties of saccharin under an investigation suggested by, and carried on under the direction of, Remsen, stole full credit for himself.)

166. Cannizzaro's determination of atomic weights was fundamental to development of the periodic law and determination of correct chemical formulas.

167. Artificial gold is produced mainly by neutron bombardment of mercury. Though the product is radioactive and cannot be used for monetary purposes or jewelry, it serves mankind beneficially in medicine.

168. Leyden Papyrus X – recipe collection for preparing false silver and gold; Stockholm Papyrus – recipe collection for preparing false gems.

169. *Auguste* Piccard, the physicist, invented an airtight gondola which he attached to a huge balloon. The aeronautical engineer and chemist, *Jean* Piccard, took a solo flight in an open gondola to test the idea of multiple balloons for stratospheric flights.

170. True. (Because of his considered radical views, Priestly was urged to flee from his Birmingham home. Once in London, he found many colleagues indifferent to his suffering and so migrated to the United States, where Thomas Jefferson befriended him.)

171. Meyer suffered almost lifelong from unbearable neuralgic pains and sleeplessness. For that reason, he took his own life.

172. True.

173. Born in Milan, Ohio, in 1847, his favorite subject was chemistry, but he was not a chemist. He did little in the way of scientific publishing and lecturing; however, the technology applied to the many things he developed and his patent disclosures provide ample evidence that he was indeed an able and successful chemist. Who was he?

174. In 1912, Paul Sabatier received the Nobel Prize for his contributions to catalysis. His use of nickel as a catalyst led to industrial applications of ______________________.

175. Johan Peter Klason wrote in his autobiography: "The forest is our happiness, our comfort and our wealth." How do these words relate to his lifelong passion?

176. Henry Edward Armstrong is not remembered so much for his contributions to experimental chemistry as he is to another aspect of chemistry. Can you name it?

177. Fritz Haber is usually the only one mentioned in the synthesis of ammonia from the elements. Yet a French chemist accomplished this composition prior to Haber. Who was he?

178. Although almost all of the eminent French chemists of the day were devoted to organic chemistry, Moissan decided upon inorganic chemistry. Can you name the subject of his first research (preparation of which had been described about fifty years earlier by Gustav Magnus)?

179. Born in Euskirchen, Germany, he was the only brother to five sisters. Though the boy preferred science, his father had made a name for himself as a successful businessman and wished his son to become his successor. Fortunately, the father agreed to allow the boy to study chemistry, for this man's work on the purines and carbohydrates won worldwide recognition and the Nobel Prize in 1902. Who was he?

180. What two capital discoveries does the name Sir William Ramsay bring to mind?

181. There are four major topics which dominated Arthur Rudolf Hantzsch's interest, but for what is he generally remembered?

173. *Thomas Alva Edison*. (The greatest inventor in history, Edison had only three months of formal education.)

174. *Hydrogenation*.

175. Klason's greatest passion was research in *lignin* (wood contains about 30 percent lignin) and he was determined to find its chemical composition. Thus his homage to trees.

176. Armstrong's great love in life was the *history of chemistry*. In his writings, he strove to arrange the subject matter of chemistry in natural periods.

177. *Henry Le Chatelier* was the first to synthesize ammonia from the elements.

178. Moissan's first inorganic research was on *pyrophoric iron*.

179. *Emil Fischer*.

180. Ramsay isolated the first rare atmospheric gas, argon; he also discovered helium, neon, krypton and xenon.

181. Hantzsch is now widely remembered as a pioneer in the application of physical methods to chemical problems. (The four topics that dominated his mind were: oximes and the steroisomerism on C=N bonds; diazo and azo compounds and the steroiosmerism on N=N bonds; the relationship between physical properties and structural properties; and the nature of acids and bases.)

182. In 1876, Otto Nikolaus Witt published his pioneering paper "Concerning the structure and formation of coloring carbon compounds", which is the foundation for the theory of ________________ and ____________________.

183. In 1905 he was appointed the first German exchange professor to the United States. The 1909 Nobel Prize in chemistry was awarded him for his work on catalysis. Can you name him?

184. It was his own infliction with the disease which led Paul Ehrlich to devise the method of staining tubercle bacilli that in essence is still in use today. True or false?

185. What is the best known name in Arabic alchemy? What controversy clouds his name still today?

186. What elements were associated with Plato's regular solids?

187. In addition to many other things, Svante Arrhenius gave us the theory of electrolytic dissociation. What is the country of Arrhenius' birth?

188. Born in Helsinki, he inherited characteristics of both the Finns and Swedes. In 1887 he started the work which was to occupy him all his life – alicyclic compounds; his stimulation was the vast naphtha reserves of Russia. Can you name him?

189. The International Committee for Atomic Weights adopted the values set by Philippe–Auguste Guye of 14.01 for nitrogen and 107.88 for silver. Yet Guye's name is inseparably connected with quite another matter. What is it?

190. In October 1892 the University of Chicago opened with funds donated by John D. Rockefeller. William Rainey Harper was its first president. The man selected to head the department of chemistry ultimately devised the theory of methylene dissociation, which postulates that a carbon compound, like an alkyl halide, first dissociates into a reactive bivalent carbon radical which can then react in various ways. Who was this Swiss–born chemist?

182. *Chromophores* and *auxochromes*.

183. *Wilhelm Ostwald*.

184. False. (When still a medical student, Ehrlich was already envisioning an art of staining microscopic preparations of both healthy and diseased tissues. He did not realize he was definitely tubercular until 1887. Incidentally, it was Ehrlich who was the founder of chemotherapy.)

185. Jabir ibn Hayyan [The first bibliography of the Muslim was completed in 987 A.D. To Jabir are ascribed hundreds of works in the alchemical section. Up to the thirteenth century, countless manuscripts bear Jabir's name (Latinized to Geber). Most historians of chemistry conclude that the famous name was used over and again by other writers in the expansion of Jabir's writings.]

186. Tethahedron – fire; cube – earth; octahedron – air; dodecahedron – heavenly element; eicosahedron – water.

187. Svante Arrhenius was born in *Sweden*. (The family name is said to be derived from that of the village Arena in the parish Malilla.)

188. *Adolph Ossian Aschan*.

189. Guye's name is inseparably connected with the electro–chemical synthesis of nitric acid.

190. *John Ulric Nef*.

191. WHO DID WHAT? ???
(Select answers from below)

a. Who was Fritz Pregl?

b. For what major contribution is the German Karl Remigius Fresenius remembered?

c. Why do we remember Francois Marie Raoult?

d. What was the contribution of Johann Jacob Balmer?

e. Who was Peter J. W. Debye?

f. Why do we remember the German chemist Carl Bosch?

Selection of Answers:

(1) He was a Dutch physicist and chemist who won the 1936 Nobel Prize in Chemistry for his studies of molecular structure.

(2) He is remembered for his research on the freezing point of solutions and for his studies on the lowering of vapor pressure by solutes in water and in nonaqueous solvents.

(3) He was a Swiss schoolmaster who discovered an empirical relationship existing between the four prominent lines of the hydrogen spectrum.

(4) He is the only Austrian to have received the Nobel Prize in chemistry (1923 for inventing a method of microanalyzing organic substances).

(5) He is best remembered for the development of a high pressure technique that created an industry of nitrogenous fertilizer materials.

(6) He is probably best known for his comprehensive research work on quantitative analysis.

Correct Answers to 191.

a. (4)

b. (6)

c. (2)

d. (3)

e. (5)

f. (1)

192. Crossword Puzzle

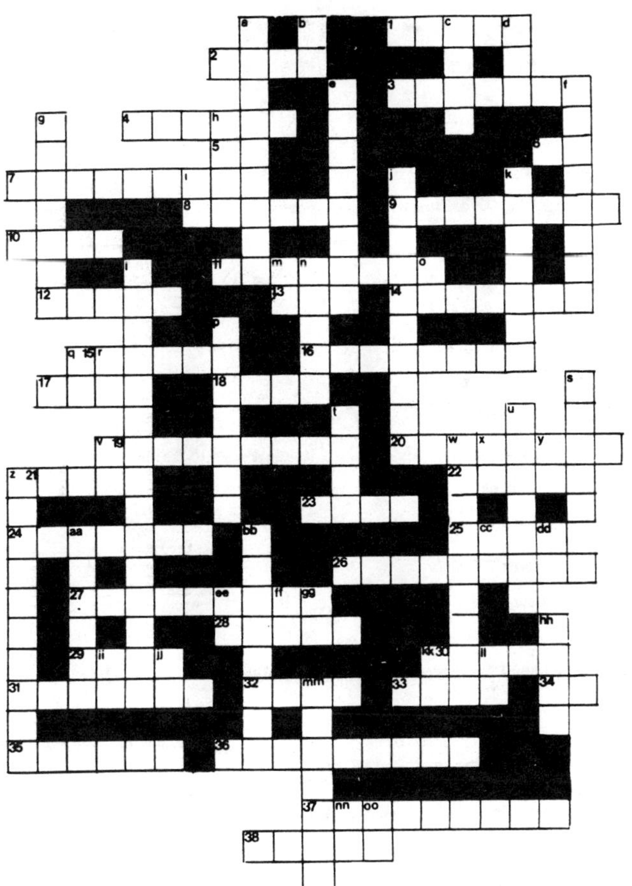

Across:

1. Pertaining to the group NH_2 united to a radical other than an acid.
2. He discovered the most elements.
3. Positive charges.
4. He proposed the atomic theory of matter in 1803.
5. _ _bonucleic acid.
6. Chemical symbol for fermium.
7. He discovered selenium and thorium.
8. In 1849, he and Regnault originated the concept of the respiratory quotient.

9. His law states that equal volumes of all gases, at the same temperature and pressure, contain same number of molecules.
10. He considered an electron gas theory of electricity.
11. An alchemist in the fifteenth century, he tried lifelong to find the seed to grow gold.
12. He studied the chemical behavior of acetylene.
13. An electrically charged atom.
14. Element No. 20.
15. Composed of two elements.
16. Its name comes from the Greek word *thallos*, meaning green twig.
17. He developed the polarimeter.
18. He studied under Silliman to become professor of Yale from 1856 to 1890.
19. Its chemical symbol is Mn.
20. The discoverer of platinum.
21. An electronic detection system.
22. The first synthetic of its kind was made by Baekeland.
23. This element falls between copper and gallium in the Periodic Table.
24. He was familiarly known as "J.J.".
25. He and Lise Meitner discovered protactinium.
26. He developed the Periodic Table.
27. A glass or pottery vessel used by alchemists (two words).
28. France's ______ Polytechnique Institute.
29. Wohler was ecstatic when he discovered he could make it without the use of kidneys.
30. She discovered polonium.
31. A diffusion process.
32. He was active in identifying degradation products.
33. An Indian physicist who invented the crescograph.
34. Chemical symbol for gold.
35. Grandson to the man who developed the theory of evolution.
36. One of the transuranium elements.
37. A spontaneous process for equalization of physical states.
38. A compound derived from ammonia.

Down:

a. His life was taken by the guillotine.
b. The chemical symbol for dysprosium.
c. Known to ancients.
d. A process developed by Roelen in 1938.

e. He devevloped the theory of relativity.

f. This element was named after the Russian engineer Samarski.

g. The first to make an organic substance from inorganic chemicals.

h. Not deviating from the essential characters of a class.

i. Chemical symbol for iridium.

j. Sixteenth century Swiss physician and chemist.

k. The chemical symbol for this element is Hf.

l. Small simple instruments used by Sir William Crookes.

m. Energy index (first letter of each word).

n. This Italian invented a machine for producing electricity.

o. Chemical symbol for sodium.

p. A compound formed by its union with water.

q. Chemical symbol for bismuth.

r. Atomic Number 53 (first two letters).

s. A certain class of organic compounds (singular).

t. A gas.

u. A German chemist, he won the Nobel prize in 1902.

v. Mass action (first letter of each word).

w. He introduced the chemistry of *ions*.

x. Light effect (first letter of each word).

y. _ _bbs is called the father of modern physical chemistry.

z. With Geiger, he invented a radiation counter.

aa. This element was discovered by Smithson Tenant in 1804.

bb. Any series of radiant energies arranged in order of wave length..

cc. Chemical symbol for aluminum.

dd. Chemical symbol for neon.

ee. Chemical symbol for helium.

ff. Alchemy origin (first letter of each word).

gg. Dalton's Law (first letter of each word).

hh. Its chemical symbol Pb comes from the Latin.

ii. He worked with Rutherford at McGill (first letter of first and last name).

jj. Activated ions (first letter of each word).

kk. Chemical symbol for cobalt.

ll. Chemical symbol for rhenium.

mm. A Russian chemist and composer.

nn. Chemical symbol for indium.

oo. Chemical symbol for iron.

Answers to Crossword Puzzle:

Across:

1. amino
2. Davy
3. protons
4. Dalton
5. Ri
6. Fm
7. Berzelius
8. Reiset
9. Avogadro
10. Albe
11. Trevisan
12. Reppe
13. ion
14. cadmium
15. binary
16. thallium
17. Biot
18. Dana
19. manganese
20. Scaliger
21. radar
22. resin
23. zinc
24. Thomson
25. Hahn
26. Mendeleev
27. Moor's head
28. Ecole
29. urea
30. Curie
31. osmosis
32. Rabe
33. Bose
34. Au
35. Darwin
36. americium
37. diffusion
38. amine

Down:

a. Lavoisier
b. Dy
c. iron
d. Einstein
e. Oxo
f. samarium
g. Woehler
h. Ir
i. True
j. Paracelsus
k. hafnium
l. spinthariscopes
m. Ei
n. Volta
o. Na
p. hydrate
q. Bi
r. Io
s. diene
t. helium
u. Fischer
v. Ma
w. Arrhenius
x. Le
y. Gi
z. Rutherford
aa. osmium
bb. spectrum
cc. Al
dd. Ne
ee. He
ff. Ao
gg. Dl
hh. lead
ii. Ro
jj. Ai
kk. Co
ll. Re
mm. Borodin
nn. In
oo. Fe

193. DO YOU KNOW???
(Select answers from below)

a. Lord Rayleigh received the 1904 Nobel Prize in Physics for studying the density of gases and for discovering argon. By what name was he born?

b. The term *entropy* was introduced to represent the degree of randomness which increases in spontaneous processes. Who coined the term?

c. *Activated molecules* is a concept introduced to describe those molecules with much greater than average energy and therefore more susceptible to reaction. Who introduced this concept?

d. A British contemporary of Lavoisier devised a water calorimeter in which he measured the heat produced by burning wax or charcoal and by a live guinea pig. Can you name him?

e. The man who invented the stethoscope died of tuberculosis, a disease on which he had become an expert. Who was it?

f. At the age of eleven, he was carrying on a wide correspondence with geologists in such a mature manner that he was invited to deliver a lecture at the Mineralogical Club in New York City, where he lived. When he arrived at the meeting in the company of his parents, no one could believe that the speaker of the day was but a boy. Can you name this precocious child who later became an eminent physicist?

Selection of Answers:

(1) J. Robert Oppenheimer
(2) Svante Arrhenius
(3) Rene Theophile Hyacinthe Laennec
(4) Adair Crawford
(5) Rudolf Clausius
(6) John William Strutt

Correct Answers to 193.

a. (6)

b. (5)

c. (2)

d. (4)

e. (3)

f. (1)

194. Al-Razi had many publications to his credit, including twelve on alchemy. Can you judge which of the following pertain to that subject? (a) *Introduction to the Theory of the Art*, (b) *Book of the Stone*, (c) *Book of the Nobility of the Art*, (d) *Book of Progression*, (e) *Book of the Secret of Secrets*, (f) *Book of the Letters to the Kings*.

195. Arthur Dehon Little's first and lasting chemical interest was cellulose. Why did he exercise such influence among the general public in his day?

196. The first commercial operation of an electrochemical process in the United States occurred during the period December 9, 1880, and January 2, 1891. Who was involved and under what circumstances?

197. When eighteen years of age, he submitted for criticism the results of his first independent chemical investigation. At the age of twenty-six, he established his coordination theory on the arrangement of atoms, for which he was awarded the Nobel Prize in 1913. Can you name him?

198. What role did Alexander Abramovich Voskresenskii play in Russian history of chemistry?

199. What were Michael Faraday's contributions to the development of the electrolytic solution theory?

200. In a friendship that spanned better than 35 years, an Armenian and a German gained world reknown in chemical research for their adopted Italy. Together they discovered a valuable antiseptic; developed a pyrolle-pyridine ring expansion; added contributions to terpene, essential oil, and pyrrole chemistry; and virtually initiated the field of organic photochemistry. Who were they?

201. The discovery of the last member of the noble gas family led to the overthrow of Dalton's principle that atoms are immutable. True or false?

202. What discovery isolated and characterized the first family of natural radioactive isotopes?

194. Each title relates to alchemy.

195. Because of Little's thought–provoking articles and addresses, which were reprinted and widely read.

196. *Herbert Henry Dow*. He separated bromine from raw brine, not by chemicals, but by an electric current.

197. The Swiss chemist *Alfred Werner*.

198. In his long career as a teacher, Voskresenskii trained many Russian chemists which earned him the nickname "Grandfather to Russian Chemists".

199. Faraday's most positive contributions in this regard are his two laws of electrolysis. His second contribution was his insight, which provided the inspiration and direction for later workers.

200. The Armenian - *Giacomo Ciamician*; the German - *Paul Silber*.

201. True.

202. The rules for radioactive decay. [Incidentally, Frederick Soddy coined the term *isotopes* to describe "...all the members of the three disintegration series, which by the constant application of these rules (for radioactive decay) fall into the same place in the periodic table, are chemically completely identical and nonseparable from one another."]

203. Lord Rayleigh (born John William Strutt) and Sir William Ramsay reported the discovery of a new element in the atmosphere in a joint paper in August 1884. In 1895 Mendeleev first heard of the discovery of the rare gas. After much reluctance, he agreed that the gases discovered by Rayleigh and Ramsay did form a new group in the periodic table. What was the first rare gas discovered and how was it isolated?

204. In 1913 a remarkable announcement was made: Using the diffusion techniques developed by Rayleigh and Ramsay, neon could be separated into light and heavy fractions with atomic weights of 20.15 and 20.28. What did this discovery prove?

205. Marie Curie is the only woman to have been awarded two Nobel Prizes. Which of the following statements are true:

(a) She was born Marie Sklodowska on November 7, in Warsaw.
(b) It was at the invitation of her sister, Bronia, that Marie went to Paris.
(c) She first met her husband Pierre at the Sorbonne.
(d) Two children were born to Pierre and Marie Curie.
(e) It was Marie who discovered radium.
(f) She received the Nobel Prize in Chemistry in 1903 for discovering radium.

206. What observation led the Curies to find the radioactive elements in pitchblende, which in turn resulted in the isolation and identification of polonium and radium?

207. While enjoying a performance of Tristan and Isolde at the state opera in Karlsruhe, Kasimir Fajans conceived the hypothesis of his "displacement laws". Can you cite these laws?

208. Lavoisier was guillotined in May 1794 for his political views. A year later, a similar fate was met by a Hungarian chemist who had boldly declared that all living things owe their being to the affinity of the atoms. Any idea who he was?

209. "Fick's laws of diffusion" and the "Fickian frame of reference" are terms which frequently appear in modern papers. Who was "Fick"?

203. Rayleigh and Ramsay discovered *argon* (from the Greek, meaning "the lazy one"). Ramsay isolated the gas by passing atmospheric nitrogen through a purification train of phosphorus pentoxide, heated copper and copper oxide, and soda–lime; the volume of uncombined gas was then passed over hot magnesium. Rayleigh prepared nitrogen from various chemical compounds which he then sparked with air enriched with oxygen over caustic solution.

204. The first definitive proof of the existence of stable isotopes.

205. Statements (a), (b), (d), and (e) are correct. Marie met Pierre at *the home of a friend*. She received the Nobel Prize (with Antoine Henri Becquerel and Pierre) in *Physics* in 1903 *for discovering radioactivity and studying uranium*.

206. The observation that the radioactivity per gram of uranium is higher in some minerals than in uranium led Mme. Curie to suspect that these minerals also contained other radioactive materials.

207. (1) Alpha particle emission is accompanied by a transition from right to left in a horizontal row of the periodic table. The observation by Soddy that this proceeds by a jump to the next but one group is assumed to be true in every case. (2) In similar fashion it is derived for beta–disintegrations, that they cause a transition to the next higher group, i.e., from left to right in a horizontal row.

208. The Hungarian was *Ignatius Martinovics* (who organized a conspiracy inspired by the ideas of the French Revolution).

209. Adolf Fick applied the concepts and methods of physics to the study of living organisms, which became the foundation of modern physiology.

210. The name of William Allen Miller seems to have slipped into obscurity. Yet he made a great contribution to stellar spectroscopy. Can you name it?

211. He had wished to become a research scientist, but he failed his third–year examinations. Still, he went on to write such imaginative works as *The Time Machine*, *The Outline of History* and *Science of Man*. Who was he?

212. Vladimir N. Ipatieff was a member of the Academy of Sciences in both Russia and the United States. Among the many honors bestowed upon him were the Lenin Prize, the Berthelot Medal in France and the Willard Gibbs Medal. What did Ipatieff accomplish that made him so highly recognized?

213. The German chemist/physicist Fritz Haber received the Nobel Prize in chemistry in 1918 for his process of synthesizing ammonia from nitrogen and hydrogen. He died in Switzerland, rejected by the country he had served so well. Why was this?

214. He was born in Russia into an Orthodox Jewish family. Named Fishel Aaronovich Levin, his family and friends called him Fedia. Although he was trained a physician, he contributed greatly to the establishment of biochemical research in the United States. Can you name him?

215. They shared the Nobel Prize in 1912 for their discovery of the reagent to synthesize organic compounds. Who were they?

216. *The Journal of the American Chemical Society* was first published under the name *The Journal of Analytical Chemistry*. Who was its first editor?

217. What was the earliest form of gravimetric analysis?

218. Who was the first to use the metric system in weighing?

219. The first convenient procedure for analyzing organic compounds was devised in 1811 by (a) Joseph Louis Gay–Lussac or (b) Louis Jacques Thenard?

210. Miller used his spectroscope to study the spectra of the celestial bodies. The spectroscope placed in juxtaposition the element being sought with the stellar spectrum. The exact coincidence of the lines proved the presence of the element.

211. The disappointed scientist was none other than British author *"H. G." (Herbert George) Wells.*

212. Under the Soviet regime, Ipatieff achieved outstanding recognition in science. Once he migrated to the U.S., he joined Universal Oil where he became known as the foremost expert in catalytic reactions in petroleum refining. Among his achievements are the invention of high pressure processes and the alkylation of paraffins with olefins.

213. Haber refused to adhere to the demands of the Nazi regime and its anti–Semitism. As a result, he retired and left his beloved Germany.

214. *Phoebus Aaron Theodor Levene.* (His name underwent many transliterations. At the time Levin was four years old, his family moved to St. Petersburg and the Hebrew name "Fishel" was changed to "Feodor". When he came to the United States, the name Feodor Aaronovich Levin somehow became transliterated as Phoebus Aaron Levene. Once he realized the proper English translation of Feodor was Theodor(e), he decided to insert Theodor as a third name.)

215. *Francois Auguste Victor Grignard* and *Paul Sabatier.*

216. *Edward Hart,* Professor of Analytical Chemistry at Lafayette College, Easton, Pennsylvania.

217. The fire assay procedure for gold and silver ores invented in Germany in the fifteenth century.

218. *Berzelius.*

219. Credit is given to both *Gay–Lussac* and *Thenard.*

220. Frank Austin Gooch developed an instrument which radically improved the handling of many kinds of precipitates. Can you name it?

221. Analytical methods based on the measurement of the weights of elements deposited on metal electrodes by electrolysis are ____________.

222. Oxalic acid and ammonium succinate were the only organic precipitants used in organic gravimetric analysis until 1885 when Ilinski and von Knorre found that alpha-nitroso-beta-naphthol could be used to determine cobalt in the presence of nickel. True or false?

223. He established the principle that when a strain is placed upon a system in equilibrium, there will be a readjustment in the direction that will most effectively relieve the strain. Who was he?

224. In his paper "On the Equilibrium of Heterogenous Substances", Josiah Willard Gibbs described the generalization now known as the *Phase Rule*. Yet its importance failed to attract attention for almost a decade. Why was this?

225. The torsion balance was of American origin. It was developed by:

(a) Edward Williams Morley (b) F. A. Roeder
(c) Alfred Springer

226. Who was editor of *Analytical Chemistry* from 1957 to 1965?

227. Freidrich August Kekule von Stradonitz coined the terms *ortho, meta* and *para* to refer to which structures?

228. By what experiments, repeated in Jean Baptiste Biot's presence, did Louis Pasteur gain this ardent supporter?

229. It was Berthelot who first introduced the term *synthesis*. Under what circumstances did this occur?

220. Gooch's invention of the *filter crucible* in 1878 was particularly suitable for precipitates that could not be ignited without decomposition and for those which were reduced by the carbon of filter paper on being ignited.

221. Analytical methods based on electrolysis are *electrogravimetric*, in which electrons act as the precipitation reagent.

222. True.

223. The French scientist *Henri Louis Le Chatelier* (the principle is commonly known as Le Chatelier's principle of mobile equilibrium).

224. It was with the development of physical chemistry that van't Hoff, van der Waals and Ostwald recognized the importance of Gibbs' *Phase Rule*. Up to that time, physicists and chemists were unable to master it because of its abstract mathematical treatment and austere style.

225, The torsion balance, developed in 1882 by *Roeder* and *Springer*, was manufactured by the Springer Torsion Balance Company.

226. *Lawrence Trenery Hallett* (whose professional career was with Eastman Kodak and General Aniline and Film Corporation).

227. The three isomeric structures of the benzene ring.

228. Using chemicals supplied by Biot, Pasteur precipitated and acidified lead salts to obtain the well-known *dextro*-rotatory acid and the hitherto unknown *levo*-rotatory acid.

229. The bi-products Berthelot observed in substituting one element for another demonstrated to him that important organic compounds can be produced from non-organic sources. The term *synthesis* was given to this process.

230. Which of the following statements are true:

(a) The ring concept was extended to naphthalene when Erlenmeyer proposed a structure of two fused rings in 1866.

(b) This was later proved incorrect by Graebe.

(c) The structure of anthracene was formulated by Graebe and Liebermann as bribenzene with alternative formulas.

(d) Phenathrene is isomeric with anthracene.

231. In the Schotten–Baumann reaction, aldol and benzoyl chloride react to form an ester. True or false?

232. In what year and under what circumstances was the official nomenclature established for organic chemistry?

233. Who was Friedrich Konrad Beilstein?

234. What caused William Augustus Tilden to turn from pharmacy to chemistry?

235. What caused the death of Belgian chemist Paulin Louyet and French chemist Jerome Nickles?

236. Edmond Fremy obtained calcium at the cathode during the electrolysis of anhydrous calcium fluoride but he was unable to isolate the gas which escaped at the anode. Moissan was successful in 1886. How was this achieved?

237. Which of the following statements are true?

(a) Amino acids are closely allied to purines.

(b) Except for a few albumins, proteins were not recognized as unique substances until well into the nineteenth century.

(c) Early studies of amino acids led to the isolation of nitrogenous compounds.

(d) The earliest amino acid isolated was cystine.

(e) Cystine was detected as a fermentation product in cheese in 1819.

(f) Glycine was first detected in gelatin.

238. In 1904 a British scientist took the Nobel Prize in chemistry. Who was he and what discovery led to this award?

230. Statements (a), (c) and (d) are accurate; (b) is not: Graebe proved Erlenmeyer's concept *correct.*

231. False. *Phenol* reacts with benzoyl chloride to form an ester.

232. In *1892 at the International Congress in Geneva.*

233. Beilstein was born in St. Petersburg of German parents. Following years of work in Germany and France, he returned to St. Petersburg to succeed Mendeleev as professor at the Imperial Technological Institute.

234. Tilden was inspired by the lectures of August Wilhelm von Hofmann. He turned to chemistry to ultimately succeed Hofmann at the Royal College of Chemistry in London.

235. The highly toxic effect of fluorine compounds was responsible for the death of Louyet and Nickles.

236. Moissan obtained fluorine gas from potassium acid fluoride dissolved in anhydrous hydrofluoric acid.

237. Statements (a), (b), (c), (d) and (f) are true; (e) *leucine*, not cystine, was detected in cheese.

238. *Sir William Ramsay* for discovering helium, neon, xenon and krypton, and for determining their place in the periodic system.

239. Alfred Werner received the Nobel Prize in 1913 for his coordination theory on the arrangement of atoms. He was born in Mulhouse, France, and served in the German army. Yet he studied, taught and died in another country. Can you name it?

240. Large-scale liquefaction of oxygen and nitrogen was achieved in 1883 by James Dewar in London and by Sigmund Florenty von Wroblevsky and Karol Stanislav Olszevski at the University of Cracow. True or false?

241. The term *osmosis* was introduced by the French physiologist Rene Joachim Henri Dutrochet. From whence does it originate?

242. This Scot's pioneering work in colloids led to his introduction of such terms as *colloids*, *crystalloids*, *dialysis* and *sols*. Can you name this chemist?

243. The ultramicroscope was developed in 1903 by H. F. W. Sideentopf and Richard Zsigmondy for what application?

244. The role of air in plant growth was suggested by Nehemiah Grew in England and Marcello Malpighi in Bologna. By what name is this phenomenon known today?

245. It was Gerardus Johannes Mulder who first coined the word *protein*. To what did he apply it and from what source comes the word?

246. The first studies on soil bacteriology were conducted by John Bennet Lawes. Where was this undertaken?

247. The work achieved with respiratory problems began with the ice calorimeter of Lavoisier and Laplace. How was this instrument used?

248. We all use the terms *Fahrenheit* and *Celsius*, yet most of us probably don't know where the terms originated. Do you?

249. Extensive research on energy metabolism was carried on by Karl von Voit and Max von Pettenkofer. What did they measure?

239. Werner's adopted country was *Switzerland.*

240. True.

241. From the Greek. (Osmosis means to push.)

242. The Scotsman in question was *Thomas Graham.*

243. The ultramicroscope made it possible to count particles and estimate their size.

244. *Photosynthesis.*

245. The term *protein* is from the Greek, meaning first substance. Mulder applied it to those nitrogenous constituents universally found in biological materials.

246. On Lawes' estate in England (known as the Rothamsted Experimental Farm).

247. The inner chamber of the ice calorimeter was surrounded by two concentric outer chambers containing ice. The outermost ice chamber insulated the inner ice chamber. The amount of heat allowed by a combustible substance of a small animal placed in the inner chamber was calculated from the amount of water collected from the inner ice chamber.

248. The Fahrenheit temperature scale was developed by the German physicist *Gabriel Daniel Fahrenheit.* The Celsius scale is the international name for the centigrade temperature scale. The name change occurred in 1948. *Anders Celsius* was a Swedish astronomer who devised the scale in 1742.

249. Using a Voit–designed respiration calorimeter, von Voit and von Pettenkofer measured oxygen consumption, carbon dioxide and water elimination, and heat production under a variety of circumstances.

250. With grandiose vision, this nineteenth century German scientist created the field of chemistry. Can you name him?

251. Edward Buchner prepared a yeast juice devoid of living cells. This enzyme was named ______________________.

252. A typhoid epidemic in London in 1905 forced the use of what method of water purification?

253. Following outbreaks of trichinosis from infected pork in 1876, the Germans organized an Imperial Health Bureau. True or false?

254. In 1810 an English best seller dealt with the adulteration of foods, drugs and household needs and gave directions for detecting various frauds. What was the nickname given to both the book and its author, the German chemist Frederick Accum?

255. In 1884 Harvey W. Wiley, director of the Bureau of Chemistry in the United States Department of Agriculture, initiated a program of systematic food analysis which still plays an important role in the establishment of reliable methods for the analysis of drugs, foods, feeds, soils, fertilizers, etc. But it was the publication of a novel which forced Congress to pass a meat inspection act and a pure food and drug act. What was this sensational piece of fiction that had such an effect on the country?

256. The Leblanc process, introduced in 1791, did not become important until the nineteenth century was well underway, yet by the end of the century its survival was threatened by the Solvay process. To what trade are both of these processes ascribed?

257. Eleuthere Irenee du Pont de Nemours, founder of The Du Pont Company, was once associated with Lavoisier. True or false?

258. Thaddeus S. Lowe, an American inventor, devised a process for alternately exposing hot coke to steam and air. What did he develop?

259. The Pittsburgh Reduction Company was the nucleus of what well-known concern of today?

250. The father of chemistry was none other than *Justus von Liebig*. (Incidentally, at the time Liebig began his studies, the study of chemistry was not possible in Germany.)

251. Buchner's yeast was named *zymase*.

252. *Chlorination*.

253. True. (It took such an outbreak to cause the country to act.)

254. Accum's book, entitled *A Treatise on the Adulteration of Foods, and Culinary Poisons*, bore a title page of a cooking vessel with a Death's Head superimposed and two serpents twined around the vessel. On the vessel was the inscription "There is Death in the Pot, 2 Kings C.IV.V.40". It was from this that both the book and the author were dubbed *Death in the Pot*.

255. *The Jungle* by Upton Sinclair. This bit of fiction so truthfully exposed meat packing houses that public outcry forced Congress to pass a meat inspection act and a pure food and drug act.

256. The Leblanc process refers to the soda process for making sulfuric acid. In 1865 the brothers Solvay developed a successful process for converting ammonium bicarbonate with salt to give sodium bicarbonate, and by 1890 Solvay process plants were operating in the United States, Russia, Germany and several other countries. (The electrolytic production of alkali and chlorine further wrecked the Leblanc industry.)

257. True. (In fact, it was the great French scientist who most often commanded young Du Pont's attention and introduced him to gunpowder.)

258. Lowe's process resulted in water gas (blau gas), which came into use as a fuel gas.

259. *The Aluminum Company of America*.

260. Although the synthetic dye industry enjoyed tremendous success in England, by 1875 the hub had become concentrated in Germany. Why was this?

261. Felix Mendelssohn is well remembered as a composer. Though of significantly less prominence, his son Paul Mendelssohn-Bartholdy none theless made a name for himself in chemistry. Any idea of how?

262. Aluminum was isolated in a highly impure form by Hans Christian Oersted of Denmark in 1825. But it was not until 1886 that two young men (one French, the other American) discovered an electrolytic method to produce the metal. Can you name these two men?

263. Fritz Rothe of Germany discovered that when nitrogen is passed over hot calcium carbide, silicon carbide is formed. True or false?

264. In 1868 Helmholtz developed a theory of vortex motion. What is this theory?

265. Joseph John Thomson set out to demolish the argument that cathode rays are not deflected by an electrostatic field. What was the logic he applied?

266. In 1880 the Scot John Aitken discovered that a fog is produced when air saturated with water vapor is cooled by a slight expansion of volume. This was confirmed in 1897 by Charles Thomson Rees Wilson, which led to the development of the Wilson cloud chamber. After 1912, this chamber was an effective tool in studies of ____________.

267. In 1900 Friedrich Ernst Dorn isolated a gaseous emanation from radium. Initially dubbed *niton*, it later was named ____________.

268. He invented the radiometer and the spinthariscope, and isolated thallium. He also invented the sodium-amalgemation process of separating gold and silver from their ores. Who was this British chemist, ranked as the foremost authority of his time on the industrial uses of chemistry?

269. Svante Arrhenius was the first of his countrymen to become chemistry's Nobel Laureate. In what year did he receive the Prize and for what discovery?

260. The English synthetic dye industry lost out to the Germans because of its inability to perceive the necessity for basic research in coal–tar chemistry. Moreover, most of those chemists connected with English dye manufacture were German and had been trained by A. W. Hofmann. When he returned to Bonn, many of his former students returned to take important positions with German producers.

261. With Carl Alexander Martius, Mendelssohn organized The Aktiengesellschaft fur Anilinfabrikation, a chemical concern more commonly known as AGFA.

262. They were *Paul L. V. Heroult of France* and *Charles Martin Hall of the United States*.

263. False. *Calcium cyanamide* is formed. (Silicon carbide was discovered by the American Edward G. Acheson.)

264. Helmholtz showed that once formed in a fluid lacking in viscosity, vortices should permanently undergo vortex motion and retain their identity.

265. Thomson reasoned that since cathode rays cause ionization of the gas in the tube, the gas attracted to the plates would neutralize the electrostatic charge and that cathode rays would therefore be deflected by an electrostatic field into a parabolic path.

266. *Radiation*.

267. *Radon*.

268. *Sir William Crookes*. (Incidentally, the electric vacuum tube he developed led to the discovery of the electron.)

269. The Swede Arrhenius received the Nobel Prize in *1903* for his *dissociation theory of ionization in electrolytes*.

270. William Henry Bragg was professor of chemistry at Leeds and later at the University of London and at the Royal Institution. With his son William Lawrence Bragg, he made it possible to establish the character of x–rays as a form of light. True or false?

271. Elemental radium was isolated in 1910 when Marie Curie and co–worker succeeded in electrolyzing fused radium chloride. What was the name of her co–worker?

272. The association of alpha–particles with helium was clearly demonstrated in 1909 by Ernest Rutherford and Thomas D. Royds. How was this accomplished?

273. The French scientist P. Villard recognized a third type of radiation, unresponsive to a magnetic field but with marked penetrating power. What was it termed?

274. It was Frederick Soddy who applied the term *isotope* to certain radioactive elements. What criteria must an element possess to be considered isotopic?

275. The instrument used by Francis William Aston to obtain a record of the particles in gas according to mass is now known as

______________________________.

276. Which of the following statements are true?

(a) The centennial of the discovery of oxygen stimulated the founding of The American Chemical Society.

(b) The Society was founded in Washington, D.C., in 1876.

(c) Established in 1879, the *Journal of the American Chemical Society* was for some years less important than Ira Remsen's *American Chemical Journal*, but the former eventually absorbed the latter.

(d) The American Chemical Society is the largest general scientific society in the world.

277. At one time, Denmark claimed two famous scientists, each with the first name "Niels" and the last initial "B" Can you name them?

278. Who was Lise Meitner?

270. True. By treating the crystal as a three-dimensional diffraction grating, the father-son team established the character of x-rays as a form of light with wavelengths ranging from a fraction of an Angstrom unit to several hundred Angstroms.

271. *Andre Debierne*. (Mme. Curie's husband had been killed in a traffic accident four years earlier.)

272. They placed radon in a glass bulb with walls so thin that only alpha-particles could pass through. This bulb was then placed inside a tube which had one end drawn into a wire-sealed capillary. By evacuating the outer tube, any gas therein was forced by mercury into the capillary. Since a control experiment using helium gas showed no evidence of a helium spectrum, it was determined that the source of helium in the first instance was the alpha-particles from the radon gas.

273. *Gamma*.

274. Isotopes are atoms of the same element that differ in atomic weight, or mass.

275. Aston used what became known as the *mass spectrograph*.

276. Statements (a) and (c) are correct. (b) The American Chemical Society was founded in *New York* in 1876. (d) The American Chemical Society is the world's *largest society dedicated to a specific discipline*.

277. Physicist *Niels Bohr* and chemist *Niels Bjerrum*.

278. Lise Meitner was an Austrian physicist who worked with Otto Hahn and Fritz Strassmann at the Kaiser Wilhelm Institute at Berlin. Her discoveries in nuclear physics played an important part in the development of atomic energy. It was she who discovered that Hahn and Strassmann's bombardment of uranium with neutrons had produced barium.

279. Black phosphorus was first discovered by P. W. Bridgman in 1914. How does it differ from white phosphorus?

280. The first notable development of platinum came with its discovery in the Ural Mountains in Russia in 1819. Today more than half the annual production of platinum results from the electrolytic recovery of nickel and copper in what country?

281. In 1930, Sr. Chandrasekhara Venkata Raman was awarded the Nobel Prize in Physics for discovering a new effect in radiation from elements. This discovery led to the widespread application of the technique which bears his name. What is the name of this technique and what is its principle?

282. It was not known in 1925 whether the proton was dislodged from the nucleus by a severe collision with an alpha-particle or whether the alpha-particle actually combined with the nucleus, which then rearranged with the ejection of a proton. Who resolved this puzzle?

283. The first atomic transmutation utilizing protons was achieved by J. D. Cockroft and E. T. S. Walton in 1931. How was this accomplished?

284. Who designed the cyclotron and in what year?

285. The Geiger counter is a familiar name to most of us. The original was designed in 1908. Geiger counters in use today are based on designs made in the 1920's. What three men were involved in the design?

286. Steel production jumped from 13,000 tons in 1860 to 4,790,320 in 1890. What product was responsible for this surge?

287. One of the first to undertake neutron bombardment of elements was a physics professor at the University of Rome. Can you name him?

288. Edwin M. McMillan and Philip H. Abelson isolated the second manmade element to be discovered. What is it called and why?

289. From whence comes *Joule's law*?

279. Black phosphorus is a dense polycrystalline polymorph, not found in nature, which resembles graphite in appearance, feel, thermal and electrical properties. Stable at room temperature in a dry atmosphere, it ignites only with difficulty and is a semiconductor. (White phosphorus, a waxy white non–conducting molecular solid of high chemical reactivity, was discovered by Henning Brand.)

280. *Canada.*

281. *Raman spectroscopy* has been of great use in the study of molecules. The Raman effect (the change of frequency of a light which is passed through a liquid or gas) provides a way to study the structure of the scattering molecules.

282. *Patrick M. S. Blackett* in England and *William D. Harkins* of Chicago by examining thousands of cloud chamber photographs.

283. Cockroft and Walton bombarded lithium oxide with protons prepared from hydrogen in a discharge tube and accelerated by a voltage multiplier.

284. The cyclotron was developed by *Ernest Orlando Lawrence* and *M. S. Livingston* at the University of California in *1931.*

285. The German physicist *Hans Geiger* and the British scientist *Sir Ernest Rutherford* developed the Geiger counter. In the 1920's, Geiger and his compatriot W. Muller redesigned the counter.

286. *Dynamite*, used to mine iron ore.

287. *Enrico Fermi.*

288. *Neptunium*, so named since it is the element beyond uranium, just as Neptune is the planet beyond Uranus.

289. James Prescott Joule, a British physicist, is credited with doing more than any other man to establish the idea that heat is a form of energy. *Joule's law* states that heat is produced in an electrical conductor.

290. On January 26, 1939, Niels Bohr and Enrico Fermi gave a verbal report at a conference on theoretical physics in Washington, D.C. What was the topic of their report?

291. This German received the Nobel Prize in chemistry three years before Fritz Haber. Like Haber, he also rejected the anti–Semitic movement. And, as did Haber, he spent his declining years in a foreign land. Can you name him?

292. A book by Nevil Vincent Sidgwick, published in 1927, brought him England's Royal Medal and international fame. What was the book?

293. What is the Diocletian legend?

294. Percy Lavon Julian, a Black American chemist, teacher and industrialist, who received his Ph.D. from the University of Vienna, developed an inexpensive method of manufacturing cortisone. True or false?

295. Kaolin, a pure white clay made of decomposed fieldspar, contains silica, alumina and water. It is widely used to make the highest grades of pottery. From what country does the word *kaolin* originate?

296. The oldest Greek manuscript dealing with Egyptian alchemy is *Marcianus 299*. In what century was it written and where is it presently kept?

(a) Fifth century; archives of Parthenon, Athens, Greece.
(b) Tenth/eleventh century; library of St. Mark's, Venice, Italy.
(c) Eighth century; St. Peter's Church, Vatican City.

297. Prior to 1887, scientists believed that a luminiferous ether existed in vacuums, through all outer space, and through all matter. Who was it that dispelled this theory?

298. In 1906 the third law of thermodynamics was developed. What is it and who was the physical chemist who introduced it?

299. In 1890, an American agricultural chemist devised a test to show the amount of butterfat in milk. Was it:

(a) Roger Ward Babson (b) Stephen Mountol Babcock
(c) Walter Bagehot

290. Bohr and Fermi reported on the *fission of uranium*.

291. *Richard Willstater*, who received the Nobel Prize in 1915 (for his research on chlorophyll), found a home in Locarno, Switzerland.

292. Sidgwick's book was entitled *Electronic Theory of Valency*.

293. Supposedly an edict of Diocletian which condemned alchemical works to flames since they were believed to be a threat to the state.

294. True.

295. *Kaolin* comes from a Chinese word meaning high hill.

296. (b) Tenth or eleventh century; it is stored in the library of St. Mark's in Venice, Italy.

297. Doubt was cast on the ether theory by American physicists Albert A. Michelson and Edward W. Morley, but it was Einstein's theory of relativity which resulted in science's change of attitude.

298. The third law of thermodynamics holds that the entropy of a pure crystalline material at absolute zero is zero. It was suggested by *Hermann Walther Nernst*.

299. (b) Babcock.

300. The world's largest lamp, which stands 42 inches high and weighs 50 pounds, uses 75,000 watts of electricity and produces as much light as 2,875 60–watt light bulbs. It was built to commemorate the 75th anniversary of Edison's invention. In what year was it built and where is it displayed?

301. In 1878 Louis Pasteur helped to prove that living organisms cause disease. In 1890, he produced antitoxin for tetanus and diphtheria. True or false?

302. A group of microscopic organisms that science has classed as both plants and animals is: (a) egleuna, (b) aeglena, (c) euglena.

303. Johannes Kepler was a German astronomer and mathematician who discovered three laws of planetary motion. Can you name them?

304. What was it Mary Stillwell and Mina Miller had in common?

305. Born an Austrian (Vienna), he studied and taught at Berlin and the University of Graz (Austria). In 1925 he received the Nobel Prize in chemistry for his method of studying colloids. What was his name?

306. August von Wassermann, a German physician and bacteriologist, is best known for ________________________________.

307. The absolute scale in centigrade units is often called the Kelvin scale. Why is this?

308. Who discovered the positron?

309. Nikolai N. Semenov, a Russian scientist who proposed theories to account for the mechanism of many complex chain reactions, shared the 1956 Nobel Prize in chemistry with Sir Cyril Hinshelwood. True or false?

310. What is the Pauli exclusion principle

311. What role did Mileva Maric play in the life of Albert Einstein?

300. The collosal lamp was erected in *1954* in *Cleveland, Ohio*.

301. False. It was *Emil von Behring* who produced the antitoxin for tetanus and diphtheria.

302. (c) euglena.

303. Kepler's three laws are: (1) Every planet follows an oval-shaped path, or orbit, around the sun; (2) an imaginary line from the center of the sun to the center of a planet sweeps out the same area in a given time; (3) the time taken by a planet to make one complete trip around the sun is its period.

304. Mary Stillwell was the first wife of Thomas Edison; Mina Miller the second.

305. *Richard Zsigmondy*. (Incidentally, it was Germany which claimed Zsigmondy at the time he received the Prize.)

306. The Wassermann blood test to determine whether a person has syphilis.

307. Because it was Lord William Thomson Kelvin, one of the great British physicists of the 1800's, who suggested the use of the gas thermometer for accurate temperature readings.

308. Proof of the existence of the positive electron was obtained by *C. D. Anderson at the California Institute of Technology in 1932*.

309. True.

310. The Pauli exclusion principle (developed by the Austrian-born, mathematical physicist Wolfgang Pauli – Nobel Prize in physics, 1945) states that no two electrons in an atom can occupy exactly the same position.

311. She was his first wife.

312. Which of the following statements are true:
(a) Glenn T. Seaborg was born in Ishpeming, Michigan, in 1912.
(b) He is a co–discoverer of the element plutonium and the fissile isotopes plutonium–239 and uranium–233.
(c) He was the first scientist to head the Atomic Energy Commission, which post he held for ten years.
(d) With Edward McMillan, he received the Nobel Prize in 1951 for his discovery of plutonium.

313. The man who had the greatest impact on modern analytical thermogravimetry was professor of microanalysis at the Sorbonne, Paris. Can you name him?

314. Gregory R. Choppin, professor of chemistry at The Florida State University, conducted post doctoral studies under Glenn Seaborg which led to the discovery of which of the transuranium elements?

315. Who is called the father of the atomic age?

316. John Tyndall, a British physicist and natural philosopher, described the action of a penicillium mold fifty years before Sir Alexander Fleming's discovery. Tyndall, however, is best known for another matter. What is it?

317. An expert in solvent extraction chemistry and former Rector of the world's largest scientific establishment, Moscow's Mendeleev Institute, he is now Minister of Education of the Soviet Union. Can you name him?

318. The bombardment of the maximum quantity of einsteinium available, in the form of the isotope einsteinium–235, with 40–Mev helium ions resulted in the discovery of which element?

319. Hanford Engineering Works in Richland, Washington (where plutonium was first produced) was built by which company?

320. The name of Thomas Midgley, Jr. is forever linked with the refrigerator. Why is this?

321. One man alone has twice received the Nobel Prize. Can you name him?

312. All the statements are true.

313. *Clement Duval* had the greatest impact on modern analytical thermogravimetry. (With co–workers, he examined the thermogravimetric properties of many hundreds of analytical precipitates.)

314. Choppin was co–discoverer of *mendelevium*.

315 *Enrico Fermi* is called the father of the atomic age (because he designed the first atomic piles and produced the first nuclear chain reaction in 1942).

316. Tyndall is best known for his experiments with light. The bluish appearance of a light beam passing through a cloudy solution is known as the "Tyndall effect".

317. *Gennadi Yagodin*.

318. *Mendelevium*.

319. *Du Pont* built Hanford.

320. Midgley developed *freon*, essential in refrigerators.

321. *Linus Pauling* (who won the 1954 Nobel Prize in chemistry and the 1963 Nobel peace prize.)

322. What is the *meson* and who discovered it?

323. Prior to the official announcement of the discovery of americium and curium, Glenn Seaborg informally revealed the existence of the two new elements on a nationally-broadcast radio program. True or false?

324. His work in the field merited the 1955 Nobel Prize and caused him to be dubbed *Mr. Photosynthesis*. Can you name him?

325. This manmade element was prepared by bombarding curium with carbon–13 ions in a cyclotron. It was isolated at the Nobel Institute for Physics in Stockholm. What is it?

326. Which of the following statements is true?
(a) Eugene Paul Wigner was born in Budapest, Hungary.
(b) He became a U.S. citizen in 1937.
(c) He won the Atomic Energy Commission's Enrico Fermi award in 1958.
(d) He shared the 1963 Nobel Prize in physics with Jensen and Mayer.

327. The gas *helium* was named for *helios,* the Greek word for sun. Similarly, *bromine*, which comes from the Greek word for stench, was named because of its odor. Where did the symbol for lead originate?

328. Lise Meitner developed a mathematical theory to explain the splitting of the uranium atom into two fragments and calculated the energy released in nuclear fission. She was also co–discoverer of one of the elements. Which one?

329. The first synthetic resin was named *bakelite*. Where does the term originate?

330. Whose words are these? "In some sort of crude sense which no vulgarity, no humor, no overstatement can quite extinguish, the physicists have known sin, and this is a knowledge which they cannot lose."

331. Jacob Volhard was probably best known because of his method for the determination of halides. True or false?

322. The meson is an *elementary nuclear particle*. It was discovered by *Hideki Yukawa* (Nobel Prize in physics, 1949).

323. True. The program was the Quiz Kids, and the next day many of the children reported to their teachers the existence of the two new elements. Seaborg in turn received a goodly number of letters from youngsters which indicated they had not been entirely successful in convincing their elders of the existence of the two new elements.

324. *Melvin Calvin*, former director of the bio–organic division of the Radiation Laboratory at the University of California–Berkeley.

325. *Nobelium* (of course!)

326. All are true statements.

327. The symbol for lead, Pb, comes from the Latin word *plumbum*, from which the English word for *plumber* is also taken.

328. Meitner (with Otto Hahn) was the discoverer of *protactinium*.

329. Bakelite was named for its developer, Leo Hendrik Baekeland.

330. The words are *J. Robert Oppenheimer's* (following detonation of the first atomic bomb in the New Mexico desert).

331. True.

332. Emil Votocek studied at the Technical Universities of Prague and Mulhouse and later worked in Gottingen and Prague. He developed a mercurimetric method that can be applied to quite strong acid solutions. Can you name it?

333. It was another Czech who suggested the term *chelometric titrations* for those volumetric methods which use chelating agents as titrants. Who was he?

334. Oxygen, the most abundant of the elements on earth, makes up about 21 percent of the air, 89 percent of the water, 65 percent of the human body, and 50 percent of the earth's crust. True or false?

335. Recipient of the Lenin Prize and the Nobel Peace Prize, in 1980 he was banished to the Soviet city of Gorki for his political views. Can you name him?

336. Who is Ylena Bonner?

337. What is the *quantum theory* and who developed it?

338. What relation was Pauline Koch to Albert Einstein?

339. Which of the following statements are true?
(a) Leo Szilard, Eugene Wigner and Edward Teller were born in Budapest, Hungary.
(b) Each became an American citizen.
(c) Enrico Fermi and Eugene Wigner originated the method which made possible the first self-sustaining nuclear reactor.
(d) It was Szilard and Wigner who influenced Einstein to encourage President Roosevelt to support atomic energy.
(e) Szilard and Teller shared the 1959 Atoms for Peace Award.

340. Who was Igor Yevgenevich Tamm and what was his relationship to Pavel A. Cherenkov and Ilya M. Frank?

341. The seventh and eighth transuranium elements, einsteinium and fermium, were discovered unexpectedly. Can you explain?

332. Votocek's mercurimetric method for the determination of chloride used sodium nitroprusside as an indicator.

333. *Rudolf Pribil* (who presently directs one of the research departments of the Czechoslovakian Academy of Sciences).

334. True.

335. *Andrei Sakharov*, the Russian physicist who gained prominence for his contribution to controlled thermonuclear reactions.

336. Bonner is Sakharov's second wife.

337. The quantum theory states that *energy, such as light, is given off and absorbed in tiny definite units called* "quanta" *or* "photons". (It was developed by the physicist *Max Planck.*)

338. Pauline Koch was Einstein's mother.

339. Statements (a), (b), and (d) are correct; (c) should read: Enrico Fermi and Leo Szilard originated the method of arranging graphite and uranium which made possible the first self-sustaining nuclear reactor; (e) should read: Szilard and *Wigner* shared the 1959 Atoms for Peace Award.

340. Tamm was a Russian physicist who shared the 1958 Nobel Prize in physics with Cherenkov and Frank.

341. Elements 99 and 100 were discovered in debris from the first large test of a thermonuclear device which took place in the Pacific on November 1, 1952.

342. The effect radiation has on living tissue was discovered accidently by Antoine Henri Becquerel. The tiny vial of radium he had carried in his vest pocket caused a severe external wound on his abdomen. What did this discovery eventually lead to?

343. Which chemical company used the slogan? "Better Things for Better Living - Through Chemistry"

344. The world's largest producer of copper is headquartered in New York City. Can you name the company?

345. In 1931 Linus Pauling presented in a paper in the *Journal of the American Chemical Society* that which he considers his single most important contribution to chemistry. Can you name it?

346. On the basis of research by the English biophysicist Maurice H. F. Wilkins, James Dewey Watson and Francis H. C. Crick devised a model of a molecular structure for which they were awarded the 1962 Nobel Prize in physiology and medicine. What did this model represent?

347. One of this century's most distinguished chemists, who devoted his life to studying the behavior of liquids and nonelectrolyte solutions, lived to be 101. Can you name him?

348. Aston introduced the term *packing fraction*. What does it mean?

349. The liberation of neutrons during fission was demonstrated by H. von Halban, F. Joliot and L. Kawarski in France and by two American groups. Who were the five comprising the American groups?

350. Gilbert Newton Lewis helped develop the modern electron theory of valence. He further introduced thermodynamics into the chemistry curriculum. At which university was he professor of chemistry and dean?

351. In 1783 a memoir was submitted to the French Academie entitled *Reflections on Phlogiston*, which openly attacked the Phlogiston theory. Who wrote this memoir? (a) Claude L. Berthollet (b) Antoine Lavoisier
(c) Antoine Francois de Fourcroy

342. Becquerel's radium wound led to the use of radiation as an agent to fight cancer.

343. It was *Du Pont's* slogan for years.

344. *Kennecott Copper.*

345. Pauling considers his most important contribution to chemistry the concept of *hybrid orbitals.*

346. The Watson–Crick model, which looks like a twisted ladder, is a model of the molecular structure of *deoxyribonucleic acid* (DNA).

347. *Joel Hildebrand*, revered professor at the University of California at Berkeley.

348. *Packing fraction* is the difference between the exact mass and the closest whole number (mass number).

349. The American groups consisted of (1) *Herbert L. Anderson, Enrico Fermi,* and *H. B. Hanstein* and (2) *Leo Szilard* and *Walter H. Zinn.*

350. The University of California at Berkeley.

351. *Reflections on Phlogiston*, often referred to as the Easter Memoir, was submitted to the French Academie by (b) *Lavoisier.*

352. Sir Humphry Davy not only invented the miner's safety lamp, but was the first person to isolate the chemical elements sodium and potassium, and the first to isolate barium, calcium, magnesium and strontium. Davy's experiments and lectures made him famous, but he also had a second calling. Can you name it?

353. The first periodic table appeared in vertical form in 1869, in which the known elements were listed in eight columns. A second, much improved table appeared two years later. How was it arranged?

354. A German is credited as an independent discoverer of the periodic law. In his *Die modernen Theorien der Chemie* he included a horizontal table in which analogous elements came under one another. Can you name this man?

355. His predecessors were Mikhail Vasilevich Lomonosov and Nikolai Nikolaevich Zinin, his contemporary Mendeleev. He invented the term "chemical structure" and went on to see its implications. His work led to discovery of the tertiary alcohols (in 1864) and the nature of tautomerism (in 1876). Who is he?

356. On Valentine's Day, 1961, Al Ghiorso of California's Lawrence Radiation Laboratory climbed a ladder to paint on the wall the discovery of a new element. What is the element?

357. Who received the first Nobel Prize in chemistry?

358. Because of his many contributions to science, Ernest Rutherford has been called "the father of nuclear science." Which of the following statements are true?

(a) Rutherford was born in Manchester, England.
(b) In 1902 he published his theory of *atomic transmutation*.
(c) For this achievement, he won the 1908 Nobel Prize in chemistry.
(d) For a time Rutherford served as professor of physics at McGill University in Montreal, Canada.
(e) He was co–discoverer of nine radioactive isotopes.

359. The American chemist, ____________________, discovered deuterium in 1931 for which he was awarded the Nobel Prize in 1934.

352. Davy had quite a reputation in the literary circles, for he was a poet and some of his verse was published by Wordsworth and Coleridge in their annual anthology.

353. Mendeleev's second table (1871) listed the elements in twelve horizontal rows, called series, so that related elements were in vertical columns, or groups, numbered I to VIII.

354. *Lothar Meyer*.

355. *Alexander Mikhailovich Butlerov*.

356. *Lawrencium* (named for Ernest O. Lawrence, the inventor of the cyclotron and founder of the Lawrence Radiation Laboratory).

357. The Dutch chemist *Jakobus Henricus van't Hoff* (for discovering laws of chemical dynamics and osmotic pressure).

358. Statements (b), (c), (d) and (e) are true; (a) Rutherford was born at Nelson, New Zealand.

359. *Harold Clayton Urey*.

360. The German physicist Walther Kossel developed the theory of ionic bonds and an American physical chemist the theory of covalent bonds. Can you name the American?

361. What is the *Uncertainty Principle*, formulated in 1925 by the German physicist Werner Heisenberg?

362. This Englishman is said to be the first to attribute the luminescence of sea water to decomposing animal matter. Can you name him?

363. The *weber* (named for the German physicist Wilhelm Eduard Weber) is:

(a) A unit of magnetic flux in the meter–kilogram–second system.

(b) The rate of work in a circuit where an electromotive force of 1 volt maintains a current of 1 ampere.

(b) The amount of electromotive force required to maintain a current of 1 ampere through a resistance of 1 ohm.

364. Hermann Kolbe, an influential organic chemist in Leipzig, denigrated van't Hoff's work which was the foundation for stereochemistry. True or false?

365. Can you identify the following tetrahedral models of van't Hoff?

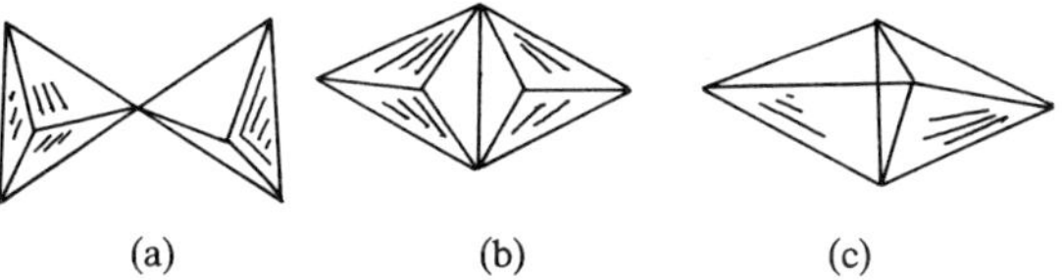

(a) (b) (c)

366. Alfred Nobel once suffered a cut to his finger which he treated with collodion (a solution of nitrocellulose in alcohol and ether). Any idea of what this invention led to?

367. The first real opportunity to study foodstuffs in the body occurred in 1822. What were the circumstances?

360. *Gilbert Newton Lewis.*

361. According to Heisenberg's Uncertainty Principle, position and momentum in the world of sub-atomic particles can never be determined simultaneously with an arbitrary degree of accuracy.

362. *John Canton.*

363. (a) *A unit of magnetic flux in the meter-kilogram-second system.*

364. True.

365. (a) ethane, (b) ethene, (c) ethyne.

366. Nobel's use of collodion led to his discovery of *blasting gelatin.*

367. The first real opportunity to study digestion came after a shooting incident at a trading post at Fort Mackinac on the United States-Canadian frontier in 1822. An Indian came out of the fray with his stomach muscles shot away and his stomach open. Though the Indian's death appeared inevitable, the post army surgeon did his best to patch up the wound and make his patient comfortable. Because the patient survived, the protruding gastric fistula allowed the surgeon to observe foodstuffs being digested at various stages.

368. Many discoveries have occurred by accident. Such was the case when an unattended mixture of anthraquinone, oxalic acid and sulfuric acid boiled to dryness, causing the anthraquinone to become sulfonated. Can you guess the result?

369. The first free radical isolated and investigated also occurred by accident. When triphenylchloromethane was heated with benzene in the absence of oxygen, a yellow, highly reactive substance resulted. This discovery opened up an entire field of study for chemists. What *was* the first free radical?

370. Better known as a physician than a chemist, this nativeborn American received his B.A. from the College of New Jersey (now Princeton) and his M.D. from the University of Edinburgh in Scotland. His signature appears on the Declaration of Independence. He is America's first professor of chemistry. Can you name him?

371. For research in extreme cold, the 1949 Nobel Prize went to:
(a) William Francis Giauque (b) William Francis Gibbon
(c) Francis Williams

372. Who was editor of the journal of *Inorganic Chemistry* from 1964 to 1968?

373. The *Doppler effect*, named for its discoverer the Austrian physicist and mathematician Christian Johann Doppler, is the apparent change in frequency of sound, light, or radio waves caused by motion. It is this effect which led to the theory that the universe is expanding. True or false?

374. The German organic chemist Kekule developed two equally valid structural formulae for benzene based on the tetravelant carbon atom. Can you show them?

375. The six noble gases were discovered in the last quarter of the nineteenth century. What are they?

376. The Italian chemist Natta Giulio was the first to produce a crystalline polypropylene in 1954. True or false?

368. The unattended mixture resulted in the discovery of the industrial synthesis of *alizarin*, superior to that obtained from madder root.

369. The first free radical isolated was *triphenylmethyl* (discovered by Moses Gomberg).

370. *Benjamin Rush*, who at the age of 23 became the first professor of chemistry in the United States.

371. *William Francis Giauque* took the 1949 Nobel Prize.

372. *Edward L. King*, Professor of Chemistry, University of Colorado, Boulder.

373. True.

374.

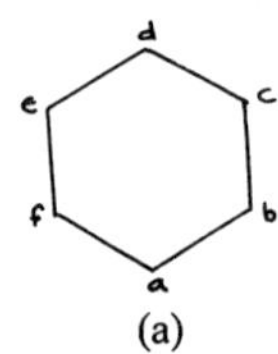

(a)

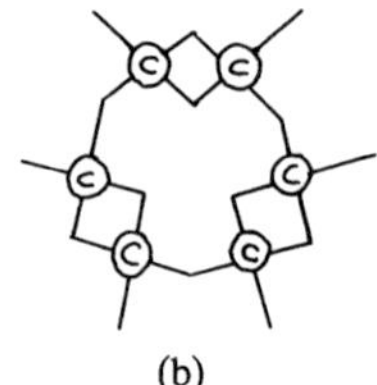

(b)

375. Helium, neon, argon, krypton, xenon and radon.

376. True.

377. Sixty years after its discovery, the first of the noble gases was synthesized by Neil Bartlett, a Canadian chemist working in the United States. This was accomplished by bonding xenon with what fluoride?

378. Before 1926, pernicious anaemia could not be cured. Then two American physicians (George Minot and William Murphy) were able to cure it by prescribing a diet of raw beef liver. Still, the search for the anaemia-curing substance continued for another twenty-two years. What was this mysterious substance that eluded science for so long?

379. For her efforts to unravel the above riddle, a British scientist took the Nobel Prize in chemistry in 1964. Can you name her?

380. The physicists M. Laue, W. Friedrich and P. Knipping discovered x-ray diffraction in crystals. This discovery led to the development of ______________________, a process whereby mathematical methods permit the conclusion as to the arrangement of atoms or ions in a crystal when they are applied to the direction and intensity of the diffracted x-rays.

381. The term *macromolecules* (Greek: makros - large) was coined by Nobel Laureate Hermann Staudinger. What does it mean?

382. Today's most commonly manufactured plastic was discovered by the exchange of one atom for another in ethylene. What are the two atoms exchanged and what is the resulting monomer?

383. Why do we remember the German chemist Karl-Dietrick Harries?

384. The Du Pont Company invested twenty-seven million dollars in the development of this substance which later resulted in many products, some very pleasing to women. What is the substance?

385. The polymerization of vinyl cyanide produced a synthetic fiber known today by the trade name *orlon*. What is this fiber?

386. Discovered by the Austrian Carl Auer von Welsbach in 1885, this element is used chiefly to prepare pigments for ceramics and ceramic glazes. Its name, which comes from the Greek, means green twin. Can you name it?

377. **Platinum(VI) fluoride, PtF_6.**

378. Vitamin B_{12}.

379. *Dorothy Mary Crowfoot–Hodgkin.*

380. X–ray structural analysis.

381. The term *macromolecules* refers to molecules which consist of at least several hundreds of atoms and which are the result of covalent bonding.

382. The exchange of a hydrogen atom in ethylene for a chlorine atom gives us the monomer of *polyvinyl chloride* (PVC).

383. Harries discovered that the basic constituent of rubber is *isoprene*.

384. Nylon.

385. Acrylic.

386. The rare–earth element *praseodymium*.

387. Basic research in polyurethane chemisty was done by Otto Bayer. Which of the following statements are true?

(a) Bayer prepared linear polymers by the condensation of diisocyanates (such as toluene diisocyanate) and dihydric alcohols (such as ethylene glycol).
(b) Mass production of polyurethanes was begun in the early 1950's.
(c) It is possible to synthesize both linear and branch polyurethanes.
(d) A linear, thermosetting polyurethane can be synthesized from buta-1, 4-diol and hexamethylene-1,6-diisocyanate.
(e) Polyurethane films have been used for such diverse items as shoe uppers and binders for rocket propellants.

388. With Emil Fischer, a biochemist hypothesized that proteins consist of amino-acid chains which are connected in an acid-amide sequence. What was the name of Fischer's co-worker?

389. *Proteins*, a term coined by Berzelius, are formed through polycondensation of alpha-amino acids. True or false?

390. In 1955, the British chemist Frederick Sanger reported the first complete analysis of the primary structure of a natural protein. What was the protein he reported?

391. In 1962, John C. Kendrew, British biochemist, and Max Ferdinand Perutz, a British-Austrian physicist, received the Nobel Prize in chemistry. What constituted their receipt of this award?

392. He studied ornithology in his undergraduate years. After spending some time with the biochemist Kalckar in Copenhagen, he went on to England's Cavendish Laboratory at Cambridge University (then headed by one of the founders of crystallography, Sir Lawrence Bragg). Though he was ignorant of Bragg's Law, with co-workers he went on to solve the puzzle of the three-dimensional structure of DNA. Can you name him?

393. In 1945, three British scientists shared the Nobel Prize for physiology or medicine for the discovery of penicillin. They were Sir Howard Walter Florey, pathologist; Sir Alexander Fleming, bacteriologist; and ______________________________, biochemist.

387. All statements are true.

388. Franz Hofmeister.

389. True.

390. The pancreatic hormone, *insulin*.

391. Kendrew and Perutz shared the prize for their work in x-ray techniques whereby they traced the structure of hemoglobin and myoglobin, two proteins found in the blood and muscles of humans and animals.

392. *James D. Watson*.

393. *Ernst B. Chain*.

394. Today's technical production of urea as a fodder additive is not based on Wohler's synthesis but on a synthesis developed by Alexander von Basarov. True or false?

395. The *Peltier effect,* which is the absorption of heat by a junction of two different metals when an electric current is passed through the junction, was named for its discoverer, Jean Charles Athanase Peltier. What was his vocation?

396. This Danish physicist and chemist laid the foundation for the science of electromagnetism. He is also credited with producing the first aluminum. Furthermore, the unit of intensity of a magnetic field is named for him. Can you name him?

397. The element nickel was discovered by the Swedish scientist Alex Cronstedt in 1751; copper was known to ancients. Today, in the United States and in Canada, the five-cent piece made of copper-nickel alloy is called a "nickel". Where does this term come from?

398. Which of the following statements are true?
(a) Paul Adrien Maurice Dirac was a French physicist.
(b) With Erwin Schrodinger, an Austrian physicist, he developed the science of wave mechanics.
(c) Dirac received degrees from Bristol and Cambridge universities.
(d) It was Dirac who predicted the presence of a positively charged electron.

399. A pioneer in chromatography, David Talbot Day was better known as a chemist than as a geologist and mining engineer. True or false?

400. Mikhail Tswett, of Russian-Italian parents, received his doctorate in science at the University of Geneva (Switzerland) in 1896. Why is he remembered today?

401. The modern thermos flask, developed initially as an insulating container for a calorimeter, was named after its Scottish inventor. Can you name him?

394. True.

395. Peltier was a *watchmaker*.

396. *Hans Christian Oersted*.

397. From *kupfernickel*: kupfer meaning copper and nickel meaning demon.

398. Statements (b), (c) and (d) are correct; (a) Direc was a *British* physicist.

399. False. The statement should read: *Day was better known as a geologist and mining engineer than as a chemist*.

400. Tswett was interested mainly in investigating chloroplast pigments. He found that when a petroleum ether solution of the pigments is filtered through a column of absorbent, the pigments are separated into colored zones. In writing about his work, he referred to the technique as the chromatographic method. The technique he developed is widely used today for both preparative and analytical purposes.

401. *Sir James Dewar*.

402. This unit of radioactivity (named for its discoverer), which represents the quantity of any radioactive species in which 3.7 x 10 to the 10th disintegrations occur per second, is called a__________.

403. The name for the rare-earth metal *thulium*, discovered by the Swedish chemist and geologist Per Teodor Cleve, originates from the Latin. What does it mean?

404. Radiocarbon dating, a technique developed by the American physical chemist Willard F. Libby, is used for what purpose?

405. In 1941, the alchemists' dream was realized with the artificial production of gold from mercury. Who were the three men responsible for this discovery?

406. What isotope was adopted as the official standard for atomic weights in 1962?

407. Promethium was isolated by Jack A. Marinsky, Lawrence E. Glendenin and Charles D. Coryell in 1945. From whence comes it name?

408. The existence of a relatively small particle having a rest mass of zero and a spin of one-half was postulated by Wolfgang Pauli in 1927 to account for the apparent lack of conservation of energy when a beta particle is emitted from radioactive nucleus. What is this small particle called?

409. The *maser*, a device invented in 1953 by the American physicist Charles H. Townes and associates, generates microwave energy by exciting molecules of certain substances and then utilizing the resulting emission as an energy source. True or false?

410. The term *mach number* represents the ratio of the speed of an object to the speed of sound relative to the medium in which the object is traveling. Where did it get its name?

411. A *muon*, an unstable electron 207 times heavier than the normal electron, is part of the lepton family of subatomic particles. When was the muon discovered?

402. *Curie.*

403. The name thulium comes form the Latin word *Thule*, land thought to be the northernmost outpost of the habitable world.

404. Radiocarbon dating determines, by measurement of the radioactive strength of their carbon–14 content, the age of carbonaceous materials.

405. *Sherr, Bainbridge* and *Anderson*.

406. Carbon–12.

407. The name is based upon the Greek myth in which Prometheus, the Titan, stole fire from the gods for the benefit of man. The name was suggested by Grace Mary Coryell, wife of Charles Coryell.

408. *Neutrino*. (Incidentally, it differs from a photon chiefly in the value of its spin.)

409. True.

410. The term came from its originator, the Austrian physicist *Ernst Mach*.

411. Muons (or *mu–mesons*) were discovered in 1936.

412. Crossword Puzzle

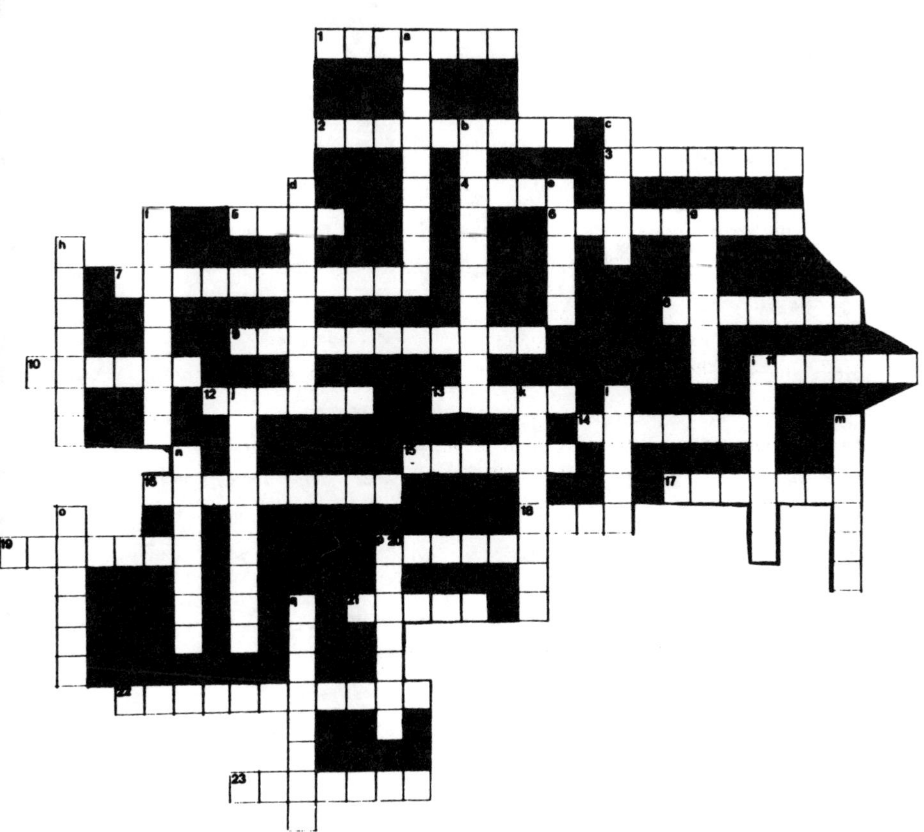

The eleven transuranium elements and their twenty-nine discoverers are listed below. From the clues given, can you properly fit them into the puzzle? The elements are: americium, berkelium, californium, curium, einsteinium, fermium, lawrencium, mendelevium, neptunium, nobelium, and plutonium. The names of the discoverers or co-discoverers are: Abelson, Browne, Choppin, Diamond, Fields, Fried, Ghiorso, Harvey, Higgins, Hirsch, Huizenga, James, Kennedy, Larsh, Latimer, Manning, McMillan, Mech, Morgan, Pyle, Seaborg, Sikkeland, Smith, Spence, Street, Studier, Thompson, Wahl and Walton.

Across:

1. He has been an advisor to every U.S. President since Roosevelt.
2. With Ghiorso, Larsh and Latimer, he discovered lawrencium.
3. Not the son of Cain, but of Abel.
4. Pictures are usually hung on it.
5. The first four letters of *mechanic*.
7. Named after one of the greatest physicists of all times.
8. It has many facets.
9. Named after the sunshine state.
10. Rhymes with *towne*.
11. The German word for *stag*.
12. *Pence* with an *s*.
13. There's one in every phone book.
14. Rhymes with tanning.
15. Several famous Englishmen bear this surname.
16. Named for the planet Neptune.
17. A priest of the same name was burned at the stake in England.
18. An arrangement of fissionable material designed to produce a chain reaction is called a ____.
19. The famous Polish composer with two *p*'s.
20. Rhymes with *shields*.
21. Some like theirs scrambled, others ______.
22. Named after the originator of the Periodic Table.
23. One who studies could be called a __________.

Down:

a. A site of the University of California.
b. This element has atomic number 103.
c. One of Christ's disciples.
d. With Abelson, he discovered the first trans-uranium element.
e. Rhymes with marsh.
f. One who suggested the name *einsteinium*.
g. Named for the discoverer of polonium.
h. His first name is Albert.
i. He was the professor in *My Fair Lady*.
j. Named for the planet Pluto.
k. Count Rumford was born with the same family name.
l. There was once a movie by the same name which starred Jimmy Stewart.
m. A paved road.
n. Shared his name with a former U.S. President.
o. The name of a family of great American bankers.
p. It was found in radioactive debris.
q. Named for a famous Swedish chemist.

Answers to Puzzle:

Across:

1. Seaborg
2. Sikkeland
3. Abelson
4. Wahl
5. Mech
6. americium
7. einsteinium
8. Diamond
9. californium
10. Browne
11. Hirsch
12. Spence
13. Smith
14. Manning
15. Walton
16. neptunium
17. Latimer
18. Pyle
19. Choppin
20. Fields
21. Fried
22. mendelevium
23. Studier

Down:

a. berkelium
b. lawrencium
c. James
d. McMillan
e. Larsh
f. Huizenga
g. curium
h. Ghiorso
i. Higgins
j. plutonium
k. Thompson
l. Harvey
m. Street
n. Kennedy
o. Morgan
p. fermium
q. nobelium

413. In the mid–1950's another exotic group of subatomic particles turned up. American scientists Murray Gell–Mann and George Zweig independently and K. Nishgima of Japan postulated the existence of these "odd" particles. Gell–Mann chose their name from a line in *Finnegans Wake*. What are these particles called?

414. Scientists have not yet determined the precise properties of the *tau* (a subatomic particle 3,490 times as heavy as an electron) which was discovered in 1975. True or false?

415. In 1969, Ghiorso announced the discovery of an element by the bombardment of californium–249 with carbon–12 nuclei in the Heavy Ion Linear Accelerator. Although the name rutherfordium was suggested by the Berkeley team, to date Element 104 remains unnamed. Why is this?

416. What is the *Milankovich cycle*?

417. Henry Gwyn–Jeffreys Moseley, a British physicist, discovered a systematic relationship between x–ray spectra and the atomic number of the elements emitting the x–rays which allowed scientists to determine the atomic number of unknown elements and to properly place them in the Periodic Table. Born in 1887, Moseley died at the age of twenty–eight. What caused his death?

418. What relation was Grace Anna Ball to Herbert Dow, founder of the Dow Chemical Company?

419. Invented in 1796 by Allesandro Volta, it was the first source of steady, direct–current electricity. What is it?

420. The first plastic to be developed (by the American John Wesley Hyatt) was celluloid. Its invention came about to replace elephant tusks, which were in short supply. What was the product Hyatt invented?

421. This man shared the Nobel Prize in chemistry in 1980 for developing compounds that are capable of producing chemical bonds useful in the manufacture of drugs. His initials are the same as the three elements with which he worked. Can you name him?

413. *Quarks.*

414. True.

415. In 1964, a group of Russian scientists under the direction of G. N. Flerov, announced the isolation of Element No. 104 and tentatively named it *kurchatovium*, in honor of the Russian nuclear physicist Igor Kurchatov. Ghiorso and team failed to find the new element by Flerov's method, but were able to obtain Element No. 104 by the bombardment of californium–249 with carbon–12 nuclei.

416. In the 1920's, Milutin Milankovich, a Serbian physicist, suggested that, because of earth's orbit and the tilt of its axis, it periodically picks up more heat from the Sun.

417. Moseley was killed in the battle of Gallipoli.

418. She was his wife.

419. The *Voltaic cell.*

420. A plastic billiard ball.

421. *Herbert Charles Brown* – H. (hydrogen) C. (carbon) B. (boron).

422. Named for a Scottish physicist of the same name, the rotation of the plane of polarization of a beam of polarized light when acted on by a magnetic field is known as the ____________________.

423. During the first half of this century, two American brothers made important contributions to atomic science. Who were they?

424. Two Czechoslovakian-born scientists shared the 1947 Nobel Prize for physiology or medicine for their work on insulin. What were their names and relationship?

425. Georg von Hevesy won the 1943 Nobel Prize for chemistry. Which of the following statements is true:

(a) He received the award for his work on the use of isotopes as indicators.
(b) He was born in Budapest, Hungary.
(c) He helped discover the element hafnium.
(d) He did his research work at Washington University, St. Louis.

426. Why do we remember the Czech scientist Jaroslav Heyrovsky?

427. The term *chelate* was introduced in 1920 by Morgan and Drew for coordination compounds whose central metal ion is attached to an organic molecule by two or more bonds. The term comes from the Greek. Any idea of its meaning?

428. The pioneers in geochemistry were American chemist Frank W. Clarke, German-born V. M. Goldschmidt, and a Russian scientist. Who was the Russian, who developed an important center of geochemical research in Moscow?

429. The term *conformation* was introduced in 1929 to apply to the arrangement of the atoms of a given molecule in space. Who introduced the term?

430. The term *vitamine* was proposed by Casimir Funk, a Polish biochemist working in London. What does it mean?

422. *Kerr effect.*

423. The Compton brothers: *Karl Taylor* and *Arthur Holly*, Nobel Physics Prize in 1927 for the *Compton effect.*

424. *Carl Ferdinand Cori* and *Gerty Theresa Radnitz Cori*, husband and wife.

425. Statements (a), (b) and (c) are correct; (d) he did his research work at the Institute of Theoretical Physics in Copenhagen.

426. Heyrovsky won the Nobel Prize in chemistry in 1959 for his invention of *polarography*, an electrochemical method of analyzing complicated chemical solutions.

427. *Chelate* comes from the Greek word *chela*, which refers to the claw of a crab or lobster.

428. *V. I. Vernadsky.*

429. *W. N. Haworth* introduced the term (in his *The Constitution of Sugars.*)

430. *Vitamine* means amine essential to life. (Funk had been trying to isolate an anti–beriberi factor from rice hulls, which he thought belonged to a class of amines essential to life. The "e" was subsequently dropped and the word *vitamin* came to be used to cover all substances essential for growth and normal development.)

431. Not only was the transuranium element *fermium* named for Enrico Fermi, but also *fermions*. What are *fermions*?

432. The Nobel Prize in chemistry in 1966 was given for the development of the *molecular–orbital* theory of chemical structure. Who was the recipient?

433. Which of the following statements are true:

(a) Paul J. Flory was chemistry's 1975 Nobel winner.

(b) He was one of the researchers that developed nylon.

(c) He received the Nobel Prize for his work in polymer chemistry, begun more than 30 years earlier at the Du Pont Experimental Station in Wilmington, Delaware.

(d) At the time of his award, Flory was professor of chemistry at Stanford University in California.

434. Herman Mark, director of high polymer research at I. G. Farben, was instrumental in the production of polystyrene plastics, the polyvinyls, the polyacrylics, and the buna–type rubbers. True or false?

435. Karl Ferdinand Braun shared the 1909 physics Nobel Prize with Guglielmo Marconi. Braun's discovery that certain crystals have the property of rectifying alternating current led to the development of

____________________________.

436. The *pion* is a type of meson assumed to be responsible for internuclear forces (sometimes referred to as the "messenger" particle). Three types of pions were discovered in 1947 by the British physicist C. F. Powell and associates. By what other name is the *pion* known?

437. Hydrogen (discovered by Cavendish in 1766) has three isotopes: (a) light hydrogen, ______________; (b) heavy hydrogen, _______________; and (c) heavy heavy hydrogen, ______________.

438. When we Americans think of "hertz", we generally think of Hertz Rent–A–Car. To a German, "hertz" means heart. Physics remembers Hertz (Gustav) as recipient of the 1925 Nobel Prize. For what work was he rewarded?

431. *Fermions* are subatomic particles with spin quantum number one–half.

432. *Robert S. Mulliken.*

433. All statements are correct.

434. True.

435. Crystal sets.

436. *Pi–meson.*

437. (a) *protium*, (b) *deuterium*, (c) *tritium*.

438. Hertz shared the Prize for proving the validity of Niels Bohr's theory of the atom.

439. It was another Hertz who gave the term *hertz* to a unit of wave frequency. Can you name him?

440. What British prime minister was trained as a chemist at Oxford?

441. The *farad*, a unit of capacitance used in electrical circuitry, was named after the great Michael Faraday. For practical applications in electronics, the farad is too large a unit. What units are more commonly used?

442. The German chemist Karl Remigius Fresenius, founder of a chemical laboratory and the journal *Zeitschrift fur analytische Chemie*, left quite a legacy. Any idea of what it is?

443. One man won the Nobel Prize twice for work in the same field. Was it? (a) Leon N. Cooper, (b) John R. Schrieffer, (c) John Bardeen.

444. A father and son shared the Nobel Prize in physics in 1915. Can you name them?

445. A mother and daughter were also Nobel winners (1903, physics; 1935, chemistry). Who were they?

446. A second father and son were physics' Nobel recipients; one in 1922, the other in 1975. Can you recall this famous family?

447. Four persons have been twice honored as Nobel Laureates. Who are they?

448. A "heavy" atomic particle, known as a *baryon* (discovered in the early 1950's), refers to nucleons and their excited states. *Baryons* include protons, neutrons, and hyperons. Can you explain the *law of conservation of baryons*?

449. *Orlon*, trademark name for a widely used synthetic fiber, was first produced in 1950. What company developed it?

439. *Heinrich Rudolph Hertz*, a German physicist.

440. *Margaret Hilda Thatcher*.

441. More commonly used are the *microfarad*, which equals one millionth of a farad; and the *micromicrofarad*, which equals one million-millionth of a farad.

442. Fresenius' two sons, R. Heinrich and Theodor Wilhelm, were competent analysts who took over the laboratory their father created. Their own sons, Ludwig and Remigius, assumed management of the laboratory in the 1920's. Though the laboratory was badly damaged in a bombing raid in 1945, it was restored and operations resumed under the management of Remigius and Ludwig's only son Wilhelm. Since the journal's creation, it has been published by the laboratory and the Fresenius family.

443. (c) It was with the other two men, Cooper and Schrieffer, that Bardeen shared the Prize in 1972 for work on superconductivity. Bardeen was originally rewarded in 1956 as co-inventor of the transistor.

444. The *Braggs*: *Sir William Henry* was the father; *Sir William Lawrence*, the son (for research on the structure of crystals by means of x-rays).

445. *Marie Curie* and her daughter, *Irene Joliot-Curie*.

446. *Niels Bohr* (1922) and *Aage N. Bohr* (1975).

447. *Bardeen* (above); *Marie Curie* (1903, physics; 1911, chemistry); *Linus Pauling* (1954, chemistry; 1962, peace); and *Frederick Sanger* (1958 and 1980, both chemistry).

448. According to the law of conservation of baryons, a proton cannot decay because there is no baryon lighter than a proton.

449. *Du Pont*.

450. Battelle Memorial Institute is one of the world's largest private research organizations. In addition to its headquarters, it has laboratories in Richland, Washington; Geneva, Switzerland; and Frankfurt, Germany. Where is it headquartered?

451. Alexander Ivanovich Oparin, a Russian biochemist, published a book called the *Origin of Life* (1936) which explains the theory he developed. What is this theory?

452. Why do we remember the American physicist Alfred Otto Carl Neir?

453. The vaccine BCG was developed in 1921 by French researchers to prevent tuberculosis. It is also used experimentally in the treatment of certain cancers. From whence came its name?

454. Antoine Henri Becquerel, who shared the 1903 Nobel Prize in physics with Pierre and Marie Curie, was the son of Antoine Cesar Becquerel. True or false?

455. In 1734, R. A. F. de Reaumur suggested that man might imitate silk. What discovery led to the eventual manufacture of *rayon*?

456. It was Carl Exel Arrhenius, an amateur mineralogist, who discovered the first rare earth (in 1787). The name is incorrect, however, since rare earths are neither rare nor earths. What are they?

457. A Spanish–American biochemist shared the 1959 Nobel Prize for physiology or medicine with Arthur Kornberg for the artificial synthesization of nucleic acids. Can you name him?

458. Rutile is a common mineral deposited on beaches. It is used to color porcelain and its refined white oxide makes the best pigment for white paint. The element it contains was named by Martin Klaproth of Germany in 1795. What is this element?

459. A *betatron*, a machine for accelerating electrons to high speed, was first built at the University of Illinois in 1940. Can you name its inventor?

450. Columbus, Ohio.

451. Oparin's theory explains how life on earth originated from chemical substance.

452. Nier developed a mass spectrograph for use in nuclear research.

453. BCG stands for *B*acillus *C*almette–*G*uerin. Its developers were Albert Calmette and Camille Guerin.

454. False. Antoine Henri Becquerel was the *grandson* of Antoine Cesar Becquerel.

455. Nitrocellulose filaments from collodion solutions.

456. Rare earths are found chiefly in the minerals monazite and bastnasite. Even the scarce rare earths (europium and letitium) are more common than the platinum–group metals.

457. *Severo Ochoa.*

458. *Titanium.*

459. *Donald W. Kerst* built the first betatron.

460. Nutrition as a science began with the French chemist Antoine Lavoisier. However, it was not until the early 1900's that biochemists were able to demonstrate the body's need for amino acids and vitamins. Lafayettte B. Mendel identified vitamin A in 1913; vitamin D was identified by Elmer V. McCollum in 1922. Were these men British or American?

461. An idea proposed in 1916 by the German physical chemist Max Bodenstein was later applied in building the first atomic bomb. Can you cite this theory?

462. Robert H. Wentorf, Jr., a physical chemist at General Electric's Research Laboratory, was the first to produce *Borazon*, which is composed of equal numbers of atoms of boron and nitrogen. What do Borazon and diamonds have in common?

463. The production of porcelain had remained a secret to Europeans for hundreds of years until 1708, when a German chemist succeeded in developing it. Who was he?

464. Walter Houser Brattain, an American physicist, shared the 1956 Nobel Prize in physics with John Bardeen and William Shockley. He was born to American parents in the city of Amoy. Any idea of where Amoy is located?

465. A center for scientific research related to energy, located on Long Island, New York, has been instrumental in the discovery of new subatomic particles. Can you name this facility?

466. Rubber was first known as *cahuchu* by South American Indians. The French called the new material *caoutchouc*. How did the name *rubber* come about?

467. Why do we remember the Polish-born, Swiss chemist Tadeus Reichstein?

468. It is added to ointments to treat skin diseases such as acne and eczema. Chemists use it to make *eosin*, a dye used in red ink. It is also important in resin adhesives. Can you name it?

460. *American.*

461. Bodenstein's theory explained the speed of certain chemical reactions in terms of "chain reactions". This he believed began by a random collision of two molecules of different chemicals which reacted and, in turn, caused another pair of molecules to react, and so on.

462. The artificially produced crystal *Borazon* is the only substance that can scratch a diamond and vice versa.

463. *Johann Friedrich Bottger.*

464. *China.*

465. *Brookhaven National Laboratories.*

466. In 1770, the English chemist Joseph Priestly discovered that the material could be used to *rub* out pencil marks. Thus, the name *rubber.*

467. Reichstein shared the 1950 Nobel Prize for physiology or medicine for his research on hormones. In 1936 he isolated cortisone and in 1933 he synthesized ascorbic acid.

468. *Resorcinol*, also known as metadihydroxybenzene.

469. Sir Robert Robinson, a British organic chemist, received the 1947 Nobel Prize in chemistry. Why?

470. In 1955, the United Nations sponsored an international conference on the peaceful uses of atomic energy at Geneva, Switzerland. Two years later, an atoms-for-peace agency was established. What is it and where is it headquartered?

471. Another specialized UN agency, commonly known by the acronym UNESCO, was established in 1946 to educate, spread culture and increase scientific knowledge. What does UNESCO stand for?

472. Instrumental in creating the National Science Foundation, this 1932 Nobel Laureate in 1945 warned, "in ten or twenty years from now, the Russians may be far ahead of us" (in science). He believed the future of mankind depended upon international peace guaranteed by a strong United Nations organization. Can you name this man?

473. It should not be forgotten that the son of Sir J. J. Thomson (1906 Nobel Prize in physics) was also physics' 1937 Nobel Laureate. For what discovery did George Thomson share the Prize with America's Clinton Davisson?

474. Prince Louis V. de Broglie received the Nobel Prize in physics in 1929. Which of the following statements are true?

(a) One of his direct ancestors served as Lafayette's chief lieutenant during the American Revolution.
(b) At thirty-two, de Broglie was admitted to the French Academy of Sciences.
(c) His older brother, Maurice, for two decades had made notable contributions in the fields of x-rays, electrons and electricity.
(d) In 1924, he suggested that the electron was composed of a group of waves which guided its path.
(e) It was for his discovery of the wave character of electrons that he shared the Nobel Prize.

475. What was the familial relationship between the Germans Hans Fischer (biochemist), Emil Fischer (chemist), Franz Fischer (electrochemist), and Hermann O. L. Fischer (organic chemist)?

469. For his research on biologically significant plant substances.

470. The *International Atomic Energy Agency*, headquartered in Vienna, Austria.

471. UNESCO stands for *United Nations Educational, Scientific and Cultural Organization*, headquartered in Paris.

472. *Irving Langmuir*.

473. For discovering the diffraction of electrons by crystals.

474. Each statement is correct.

475. The only familial relationship existing among these Fischers was between Emil Fischer (father) and Hermann O. L. Fischer (son).

476. Julian Seymour Schwinger, Richard P. Feynman and Sin–itiro Tomonago shared the 1965 Nobel Prize in physics for development of an improved theory of quantum electrodynamics. What does this theory make possible?

477. Sir Francis Galton was a British scientist who became known for research in meteorology, heredity and anthropology. What was his relationship to Charles Darwin?

478. Glauber's Salt has been used as a laxative since the 1600's. For whom was it named?

479. Willis Eugene Lamb, Jr. and Polykarp Kusch shared the 1955 Nobel Prize in physics. Lamb was awarded for discoveries on the structure of the hydrogen spectrum. For what work was Kusch cited?

480. Feodor Lynen, a German chemist, shared the 1964 Nobel Prize for physiology or medicine with an American biochemist. What was his name?

481. In 1839, Benjamin Franklin Goodrich discovered the process of vulcanizing rubber. True or false?

482. What metal, discovered by the French scientist Boisbaudran in 1875, will melt in your hand?

483. Aspartame, an artificial sweetener, was approved by the Food and Drug Administration in July 1983 for use in soft drinks. By what brand names is it known?

484. A well–known chemistry rule states that carbon can form no more than four bonds in stable compounds. Yet in September 1983, an international team of chemists presented evidence for a carbon atom with more than four atoms surrounding it. How is this possible?

485. England's multi–talented Charles Percy Snow was a well–known scientist, but was even more famous in another field. Can you name it?

476. The theory enables scientists to predict accurately the effects of electrically charged particles on each other in a radiation field.

477. He was his cousin.

478. For the German chemist *Johann Rudolf Glauber*.

479. For determining the magnetic moment of the electron.

480. *Konrad E. Bloch*.

481. False. It was *Charles Goodyear* who discovered the process of vulcanizing rubber. (Incidentally, this occurred while he was in prison for debts.)

482. *Gallium*, which has a melting point of 30 degrees C.

483. *Equal* (1981) and *Nutrasweet* (1983).

484. The chemists calculated that an arrangement of one carbon atom bonded to five or six lithium atoms would be stable. Instead, only four lithium atoms bonded to carbon, and these four atoms were bonded to the one or two extra lithium atoms. Thus the rule for carbon was not broken.

485. Snow was most famous as a *novelist*.

486. In 1972, two important developments occurred in biochemistry. English scientists clarified the nature of vitamin D and its activity in the body. American scientists succeeded in making a model of a cell structure never seen clearly because of its size. What are the results of these two developments?

487. It was a British scientist who founded our Smithsonian Institution. Can you name him and the mineral named in his honor?

488. Hafnium, discovered by Dirk Coster and Georg von Hevesy in 1923, is always found with the similar and much more common element, zirconium. Is this what the name *hafnium* implies?

489. The first practical spark chamber (built in 1959), used to make visible the paths followed by electrically charged atomic particles, was designed by Japanese physicists Fukiu and Miyamoto. True or false?

490. A Russian scientist shared the 1956 Nobel Prize in chemistry with Sir Cyril Hinshelwood for work on chemical chain reactions. Can you name the Russian?

491. Lev Davidovich Landau won the Nobel Prize in physics in 1962 for research that helped explain some strange properties of liquid helium, such as its ability to creep. What was it he discovered?

492. An American physicist who helped develop the cyclotron won the Nobel Prize in physics in 1939 and the 1957 Enrico Fermi award. Was it?

(a) Willard Frank Libby
(b) Ernest Orlando Lawrence
(c) Irving Langmuir
(d) Samuel Pierpont Langley

493. Heavy water was first separated in 1932 by Harold C. Urey. What isotope does it contain that makes it heavier than ordinary water?

494. Named for the sun, this element is used in rockets, in electric arc welding, to fill balloons, and as an aid to asthmatics. Can you name it?

495. The first linear accelerator was built in 1928 by the Norwegian physicist Rolf Wideroe. What is its purpose?

486. Vitamin D was discovered to function more as a hormone than as a vitamin. The model of the cell structure, ribosome, will allow scientists to relate ribosome form to its function.

487. *James Smithson*; the mineral is *smithsonite*.

488. It is not. The word *hafnium* comes from *Hafnia*, the Latin name for Copenhagen.

489. True.

490. *Nikolai N. Semenov*.

491. Landau discovered that helium below –455.8 degrees F moves as a system of waves, not as single particles.

492. (b) Lawrence.

493. *Deuterium*.

494. *Helium*.

495. A device, sometimes called a *linac*, which propels atomic particles at high speeds.

496. In 1983, chemistry's Nobel Prize went to Henry Taube of Stanford for his discovery of the basic mechanism of chemical reactions. The Prize in physiology or medicine that year truly surprised its octegenarian recipient. Why was this?

497. He was first secretary and director of the Smithsonian Institution and later president of the National Academy of Sciences. The unit of electrical inductance is named for him. Can you name him?

498. The Swedish chemist Theodor Svedberg won the Nobel Prize in chemistry in 1926. What development made him famous?

499. The International Union of Pure and Applied Chemistry favors the Latin name for elements. The names beryllium, niobium, lutetium, hafnium, and wolfram are Latin. By what names are these elements more commonly known?

500. The elements sodium, potassium and antimony are also known by a second name. What are they?

501. Elements known to the ancients have national names which vary from country to country. Do you know the English for the German elements *Wasserstoff, Sauerstoff, Kohlenstoff* and *Stickstoff*?

502. The French scientists Auguste and Louis Jean Lumiere were brothers who became known for their work in photochemistry. What was it they invented?

503. The fields of chemistry and physics many times overlap. So it was once again in 1971 that a Canadian physicist won the Nobel Prize in chemistry for determining the electronic structure of molecules. Can you name him?

504. John Howard Northrop and Wendell Meredith Stanley would not have shared chemistry's 1946 Nobel Prize with William Graham Sumner had it not been for Sumner. What did Sumner achieve that stimulated the other two?

496. The 81-year-old recipient, *Barbara McClintock*, was rewarded for work she had conducted some *forty* years earlier on the way genes determine hereditary changes in plants and animals. (In making the award, the Nobel committee said she had been "far ahead" of others in her field. It seems to the author that the committee was "far behind" in selecting her.)

497. *Joseph Henry.*

498. The *ultracentrifuge*, which can spin materials so fast that they have 500,000 times the force of gravity acting upon them.

499. Beryllium - glucinium; niobium - columbium; lutetium - cassiopeium; hafnium - celtum; wolfram - tungsten.

500. Sodium - natrium; potassium - kalium; antimony - stibium.

501. Wasserstoff - *hydrogen*; Sauerstoff - *oxygen*; Kohlenstoff - *carbon*; Stickstoff - *nitrogen.*

502. They invented the *Cinematographie*, a combination motion-picture camera, printer and projector.

503. *Gerhard Herzberg.*

504. Sumner extracted the enzyme urease from the jack bean and converted the crude material into pure crystals. Northrop crystallized other important enzymes and Stanley crystallized the first virus.

505. Saccharin, discovered in 1879, is an artificial sweetener widely used by dieters and diabetics. What is it made from?

506. Before isolation of the tobacco mosaic virus in 1935 by biochemist Wendell Meredith Stanley, scientists had assumed that viruses were submicroscopic living organisms. What did Stanley's discovery show?

507. The discovery of sulfa drugs was also an accident. In 1908, while investigating another matter, German chemist Paul Gelmo discovered chemicals that eventually led to the sulfa drugs. But what was he looking for?

508. Why do we remember German scientists Adolf Windaus and Heinrich O. Wieland?

509. Recently the term *xenobiotic* was introducecd into the English language. How does it apply to chemistry?

510. *Taggant* is another late term applicable to chemistry. What is its meaning?

511. The heaviest of the rare–earth elements, discovered by the French scientist Georges Urbain in 1907, is named after the ancient name for Paris. What is it?

512. Nickel, discovered in 1751 by Alex Cronstedt, is often used in structural work and electroplating because of its resistance to corrosion. In which of the following applications is nickel used?

(a) To promote certain chemical reactions by catalysis.
(b) In the Edison storage battery.
(c) To electroplate printing plates.
(d) To improve the properties of cast iron and steel.

513. Bikini Atoll, formerly called the Escholtz Islands, was evacuated so that the United States could use the island as a base for atomic and hydrogen bomb tests. An interesting development in women's fashions came about because of the name change. What was it?

505. *Toluene* and *petroleum*.

506. He showed the tobacco mosaic to be a protein molecule.

507. Gelmo was looking for better dyes for woolen goods.

508. For their work with steroids. Wieland received chemistry's 1927 Nobel Prize for studying gall acids and related substances and Windaus was recipient the next year for his studies of sterols and their connection with vitamins.

509. *Xenobiotic* is a foreign substance capable of harming or affecting a living organism, or of being a xenobiotic. XENO (a computer program) can assist chemists in predicting the biological activity of xenobiotic compounds.

510. *Taggant* is a chemical substance added to another substance to aid in detection and identification.

511. *Lutetium*, from *Lutetia*, the ancient name for Paris.

512. Nickel is used in all four applications.

513. The teeny, tiny bathing suit, known as the *bikini*, was named for the Atoll after the first hydrogen bomb test.

REFERENCES

1. *The World Book Encyclopedia*, Field Enterprises Educational Corporation, Chicago, IL, 1981, and Year Books thereto.

2. *Great Chemists*, Eduard Farber (Ed.), Interscience Publishers, New York, London, 1961, Vol. I and II.

3. Hein, Morris, *Fundamentals of College Chemistry*, Fourth Edition, Dickenson Publishing Company, Inc., Encino, CA, 1977.

4. Whitehead, Don, *The Dow Story,* McGraw–Hill Book Company, New York, 1968.

5. Eidinoff, Maxwell Leigh, and Ruchlis, Hyman, *Atomics for the Milions,* McGraw–Hill Book Company, New York, 1974.

6. *A History of Analytical Chemistry*, Laitinen, Herbert A., and Ewing, Galen W. (Eds.), American Chemical Society, Washington, D.C., 1977.

7. Dutton, William S., *Du Pont, One Hundred and Forty Years,* Charles Scribner's Sons, New York, 1942.

8. Gregory, Richard, *Discovery, or The Spirit and Service of Science*, The Macmillan Company, New York, 1924.

9. Seaborg, Glenn T., *Man–Made Transuranium Elements*, Prentice–Hall, Inc., Englewood Cliffs, NJ, 1963.

10. Glasstone, Samuel, *Sourcebook on Atomic Energy*, D. Van Nostrand Company, Inc., New York, 1950.

11. Clark, Ronald W., *Einstein, The Life and Times*, Avon Books, New York, 1971.

12. Dyer, Frank Lewis, and Martin, Thomas Commerford, *Edison, His Life and Inventions*, Harper & Brothers Publishers, New York, 1910, Vols. I and II.

13. Thompson, Charles J., *Alchemy*, Sentry Press, New York, 1974.

14. *Du Pont, The Autobiography of an American Enterprise*, Charles Scribner's Sons, New York, 1952.

15. Ihde, Aaron J., *The Development of Modern Chemistry*, Harper & Row, New York, 1964.

16. Reid, Robert, *Marie Curie*, New American Library, New York, 1974.

17. Henehan, John F., *Men and Molecules*, Crown Publishers, Inc., New York, 1966.

18. Groueff, Stephane, *Manhattan Project*, Little, Brown and Company, Boston, 1967.

19. Reiche, Charles-Albert, *A History of Chemistry,* Hawthorn Books, Inc., New York, 1963.

20. Snow, G. P., *The Physicists*, Little, Brown and Company, Boston, 1981.

21. Neubauer, Alfred, *Chemistry Today: The Portrait of a Science*, Arco Publishing Inc., New York, 1983.

22. Jaffe, Bernard, *Crucibles: The Story of Chemistry*, Dover Publications, Inc., New York, 1976.

23. *The Encyclopedia of Chemistry*, Clark, George L. (Ed.-in-Chief), Van Nostrand Reinhold Company, New York, 1966.

24. Asimov, Isaac, *Asimov on Chemistry*, Doubleday & Company, Inc., Garden City, NY, 1974.

25. Tyndall, John, *Faraday as a Discoverer*, Thomas Y. Crowell Company, New York, 1961.

26. McCue, J. J. G., *The World of Atoms: An Introduction to Physical Science*, The Ronald Press Company, New York, 1956.

27. Rouze, Michael, *Robert Oppenheimer: The Man and his Theories*, Fawcett Publications, Inc., Greenwich, Conn., 1962.

28. Campbell, Murray, and Hatton, Harrison, *Herbert H. Dow: Pioneer in Creative Chemistry*, Appleton–Century Crofts, Inc., New York, 1951.

29. *The New Treasury of Science*, Shapley, Harlow, Rapport, Samuel, and Wright, Helen (Eds.), Harper & Row, New York, 1965.

30. *Who's Who in Atoms 1960*, Vallancey Press, London, 1960, Vols. I and II.

31. *American Men and Women of Science*, R. R. Bowker Company, New York and London, 1977.

APPENDIX

LEST THEY NOT BE FORGOTTEN!

Abelson, Philip Hauge (b. 1927) nuclear physicist.
Accum, Frederick (1769–1838) chemist.
Acheson, Edward G. (1856–1931) chemist, inventor.
Agricola, Georgius (or Georg Bauer) (1490/94–1555) physician, mineralogist, metallurgist, chemist, geologist, engineer.
Aitken, John (1839–1912) physicist, meteorologist.
Ainsworth, William A., manufacturer of balances.
Al-Razi (850/860–923/932) Persian physician.
Anaximandros of Miletos (610–545 B.C.) Greek mathematician, astronomer, biologist.
Anderson, C. D. (b. 1905) physicist.
Anderson, Herbert L (b. 1914) physicist.
Arago, Dominique Francois Jean (1786–1853) physicist, astronomer, mathematician.
Aristotle of Stagira (384–322 B.C.) Greek natural philosopher, naturalist.
Armstrong, Henry Edward (1848–1937) chemist.
Arrhenius, Carl Exel (1757–1824) mineralogist.
Arrhenius, Svante August (1859–1927) chemist, physicist.
Aschan, Adolph Ossian (1860–1939) chemical engineer, geologist.
Aston, Francis William (1877–1945) physicist.
Avicenna (980–1037) Arab physician.
Avogadro, Amedeo, Conte di Quaregna (1776–1856) physicist, chemist.

Babcock, Stephen Moulton (1843–1931) agriculturist, chemist.
Bacon, Francis, 1st Baron Verulam, Viscount St. Albams (1561–1616) natural philospher.
Baekeland, Leo Hendrick (1863–1944) chemist, inventor.
Baeyer, Johann Friedrich Wilhelm Adolf von (1835–1917) chemist.
Bainbridge, John (1582/83–1643) astronomer.
Balard, Antoine-Jerome (1802–1876) chemist.
Balmer, Johann Jacob (1825–1898) physicist, mathematician.
Bancroft, Wilder Dwight (1867–1953) physical chemist.
Barasov, Alexander von.
Barbier, Philippe Antoine (1848–1922).
Bardeen, John (b. 1908) electrical engineer, physicist.
Bartlett, Neil (b. 1932) chemist.
Baumann, Eugen (1846–1896) organic chemist.
Bayer, Otto (1902–1978) chemist.

Becher, Johann Joachim (1635–1682) chemist, physician.
Becquerel, Antoine Cesar (1788–1878) physicist.
Becquerel, Antoine Henri (1852–1908) physicist.
Behring, Emil von (1854–1917) bacteriologist.
Beilstein, Friedrich Konrad (1836–1906) editor, chemist.
Bendetti-Pichler, Anton Alexander (b. 1894).
Berard, Auguste (1802–1846) surgeon.
Berthelot, Marcelin Pierre Eugene (1827–1907) chemist.
Berthollet, Amedee, son of Claude-Louis.
Bertholett, Claude-Louis Comte (1748–1822) chemist.
Berzelius, Jons Jakob (1779–1848) physician, chemist.
Biot, Jean Baptiste (1774–1862) mathematician, physicist, astronomer.
Bjerrum, Neils (1879–1958) chemist.
Black, Joseph (1728–1799) chemist, physicist.
Blackett, Patrick M. S. (b. 1897) physicist.
Bloch, Konrad E. (b. 1912) biochemist.
Bodenstein, Max (1871–1942) physical chemist.
Bohr, Aage N. (b. 1922) nuclear physicist.
Bohr, Niels (1885–1962) physicist.
Boisbaudran, Paul Emile Lecoq de (1838–1912) chemist, physicist.
Bonner, Ylena (b. 1923) physician and teacher.
Borodin, Alexander (1834–1887) chemist, composer.
Bosch, Carl (1874–1940) chemist.
Bose, Sir Jagadis Chunder (1858–1937) physicist.
Bottger, Johann Friedrich (1682–1719) chemist.
Boyle, Robert (1627–1691) chemist, physicist.
Braconnot, Henri (1781–1855).
Brand, Hennig (d.c. 1692) chemist.
Bragg, William Henry (1862–1942) physicist.
Bragg, William Lawrence (b. 1890) physicist.
Brattain, Walter House (b. 1902) physicist, inventor.
Braun, Karl Ferdinand (1850–1918) physicist.
Bridgman, Percy William (1882–1961) physicist.
Broglie, Louis Victor P. R., Prince (b. 1892) physicist.
Broglie, Maurice de (brother of Louis).
Brown, Herbert Charles (b. 1912) inorganic chemist.
Browne, C. I.
Browne, Charles Albert (1870–1947) agricultural chemist.
Buchner, Edward (1860–1917).
Bunsen, Robert Wilhelm Eberard von (1811–1899) chemist.
Butlerov, Aleksander Mikhailovich (1828–1886) chemist.

Calmette, Albert Leon Charles (1863–1933) bacteriologist.
Calvin, Melvin (b. 1911) organic chemist.
Candolle, Alphonse Louis Pierre Pyrame de (1806–1893) botanist.
Cannizzaro, Stanislao (1826–1910) chemist.
Canton, John (1718–1772) physicist.
Carver, George Washington (1864?–1943) botanist.
Cavendish, Henry (1731–1810) physicist chemist.
Celsius, Anders (1701–1744) astronomer.
Chain, Ernst Boris (b. 1906) biochemist, pathologist.
Chancourtois, Alexandre Emile Beguyer de (1819–1886) geologist.
Chaptal, Jean Antoine Claude, Compte de Chanteloup (1756–1832) chemist.
Charles, Jacques Alexandre Cesar (1746–1823) physicist, areo pioneer, inventor.
Chatelier, Henry Louis Le (1850–1936) chemist, metallurgist.
Cherenkov, Pavel A. (b. 1904) nuclear physicist.
Chesebrough, Robert A. (1837–1933).
Chevreul, Michel Eugene (1786–1889) chemist.
Choppin, Gregory R. (b. 1927) nuclear, inorganic chemist.
Ciamician, Giacomo (1857–1922) chemist.
Clarke, Frank Wigglesworth (1847–1931) geologist, geochemist.
Clausius, Rudolf (1822–1888) physicist.
Cleve, Per Teodor (1840–1905) chemist.
Cockroft, Sir John Douglas (b. 1897) physicist, engineer.
Collet-Descostils.
Compton, Arthur Holly (1892–1962) physicist.
Compton, Karl Taylor (1887–1954) physicist.
Cooper, Leon N.
Cori, Carl Ferdinand (b. 1896) biochemist.
Cori, Gerty Theresa Radnitz (1896–1957) biochemist.
Coryell, Charles D.
Coster, Dirk (b. 1889) physicist, meteorologist.
Couper, Archibald Scott (1831–1892) chemist.
Crafts, James Mason (1839–1917) physicist, chemist.
Crawford, Adair (1748–1795) physician.
Crick, Francis H. C. (b. 1916) biologist.
Cronstedt, Alex Fredrik, Baron (1722–1765) chemist, metallurgist.
Crookes, Sir William (1832–1919) chemist.
Cullen, William (c. 1710–1790) chemist.
Curie, Marie (1867–1934) physical chemist.
Curie, Pierre (1859–1906) physicist, chemist.

Dana, James Dwight (1813–1895) geologist, mineralogist.
Daguerre, Louis Jacques (1789–1851) inventor.
Dalton, John (1766–1844) physicist, chemist, mathematician.
Darwin, Davisson Clinton Joseph (b. 1881) physicist.
Davy, Edmund (1785–1857) chemist.
Davy, Sir Humphry (1778–1829) chemist.
Day, David Talbot (1859–1925) geologist, mining engineer, chemist.
De Bierne, Andre Louis (b. 1874) chemist.
Debye, Peter Joseph W. (1884–1966) physicist.
Descartes, Rene (1596–1650) mathematician, natural philosopher.
Dewar, Sir James (1842–1923) chemist, physicist.
Diamond, Herbert (b. 1925) nuclear chemist.
Diocletian (296 A.D.) Greek physician.
Dirac, Paul Adrien Maurice (b. 1902) physicist.
Doppler, Christian Johann (1803–1853) physicist, mathematician.
Dorn, Friedrich Ernst (1848–1916) chemist.
Dow, Herbert Henry (1866–1930) chemist.
Dulong, Pierre Louis (1785–1838) chemist, physicist.
Dumas, Jean-Baptiste Andre (1800–1884) chemist.
Du Pont de Nemours, Eleuthere Irenee (1771–1834) printer.
Dutrochet, Rene Joachim Henri (1776–1847) physiologist, biologist.
Duval, Clement.

Edison, Thomas Alva (1847–1931) inventor.
Ehrlich, Paul (1854–1915) bacteriologist.
Einstein, Albert (1879–1955) physicist.
Epicurus of Samos (341–270 B.C.) mineralogist, geologist.
Ercker, Lazarus (1530?–1594?) assayist.
Erlenmeyer, Emil (1825–1909) chemist.

Fahlberg, Constantine (1850–1910) chemist.
Fahrenheit, Gabriel Daniel (1686–1736) chemist.
Fajans, Kasimir (b. 1887) physical chemist.
Faraday, Michael (1791–1867) chemist, physicist, inventor.
Fermi, Enrico (1901–1954) physicist.
Feynman, Richard Phillips (b. 1918) physicist.
Fick, Adolf (1829–1901) physiologist.
Fields, Paul (b. 1919) chemist.
Fischer, Emil (1852–1919) chemist.
Fischer, Franz (1877–1947) electrochemist.
Fischer, Hans (1881–1945) biochemist.
Fischer, Hermann O. L. (1888–1959) organic chemist.

Fleming, Sir Alexander (1881–1955) bacteriologist, physician.
Flerov, Georgii Nikolaevich (b. 1913) nuclear physicist.
Florey, Sir Howard Walter (1898–1968) pathologist.
Flory, Paul John (b. 1910) chemist.
Fourcroy, Antoine Francois, Comte de (1755–1809) chemist.
Fournier d'Albe, Edmund Edward (1868–1933) physicist, inventor.
Franciscus Sylvius (or Franz de la Boe; Du Bois, Francois) (1614–1672) anatomist, physician, physiologist, chemist.
Frank, Ilya Mikhailovich (b. 1908) physicist.
Frasch, Herman (1851–1914) chemist, inventor.
Fremy, Edmond (1814–1894) chemist.
Fresenius, Karl Remigius (1818–1897) chemist.
Fresenius, Ludwig (1866–1936) chemical analyst.
Fresenius, R. Heinrich (1847–1920) chemical analyst.
Fresenius, Remigius (b. 1878) chemical analyst.
Fresenius, Theodor Wilhelm (1856–1936) chemical analyst.
Fresenius, Wilhelm (b. 1913) chemical analyst.
Fried, S. M.
Friedel, Charles (1832–1899) mineralogist, chemist.
Friedrich, W, physicist.
Funk, Casimir (b. 1884) biochemist.

Gadolin, Johan (1760–1852) chemist.
Galton, Sir Francis (1822–1911) anthropologist, inventor.
Gay-Lussac, Joseph Louis (1778–1850) chemist, physicist.
Geiger, Hans (b. 1882) physicist.
Gell-Mann, Murray (b. 1929) physicist.
Gelmo, Paul.
Gerhardt, Charles Frederic (1816–1856) chemist.
Ghiorso, Albert.
Giauque, William Francis (b. 1895) chemist.
Gibbs, Josiah Willard (1839–1903) chemist, physicist.
Giulio, Natta, chemist.
Glauber, Johann Rudolf (1604–1668/70) chemist, physician.
Glendenin, Lawrence Elgin (b. 1918) nuclear chemist.
Gmelin, Christian Gottlob, chemist, pharmacist.
Gmelin, Johann Friedrich (1748–1804) pharmacist, chemical historian.
Gmelin, Leopold (1788–1853) chemist.
Goldschmidt, Victor Moritz (1888–1947) geologist.
Gomberg, Moses (1866–1947) chemist.
Gooch, Frank Austin (1952–1929) chemist.
Goodrich, Benjamin Franklin (1841–1888) army surgeon, entrepreneur.

Goodyear, Charles (1800–1860) inventor.
Graebe, Karl (1841–1927) chemist.
Graham, Thomas (1805–1869) chemist.
Grew, Nehemiah (1641–1712) botanist.
Grignard, Francois Auguste Victor (1871–1935) chemist.
Guerin, Camille.
Guldberg, Cato Maximilian (1836–1902) mathematician, chemist.
Guthrie, Samuel (1782–1848) chemist, inventor.
Guye, Philippe-Auguste (1862–1922) chemist.

Haber, Fritz (1868–1934) chemist.
Hahn, Otto (b. 1879) physical chemist.
Halban, Hans von, physicist.
Hales, Stephen (1677–1761) physiologist, inventor, botanist, chemist.
Hall, Charles Martin (1863–1914) chemist, inventor.
Hall, Lloyd, chemist.
Hallett, Lawrence Trenery (b. 1900) analytical chemist.
Hanstein, H. B., physicist.
Hantzsch, Arthur Rudolph (1847–1935) chemist.
Harkins, William D.
Hart, Edward (1854–1931) chemist.
Hartmann, Johannes (1563–1631).
Harries, Karl-Dietrick (1866–1923) chemist.
Harvey, Bernard George (b. 1919) nuclear physicist.
Hatchett, Charles (1765?–1847) chemist.
Hauy, Abbe Rene Just (1743–1822) mineralogist.
Haworth, Walter Norman (1883–1950) chemist.
Heisenberg, Werner Karl (b. 1901) physicist.
Helmholtz, Hermann Ludwig Ferdinand von (1821–1894) physiologist, anatomist, physicist.
Helmont, Jan Baptista van (1577–1644) chemist, physician.
Henry, Joseph (1797–1878) physicist, inventor.
Heroult, Paul L. T., scientist.
Hertz, Gustav (1887–1950) physicist.
Hertz, Heinrich Rudolph (1857–1894) physicist, inventor.
Herzberg, Gerhard (b. 1904) physicist.
Hesiod (700's B.C.) philosopher.
Hevesy, Georg von (b. 1885) chemist.
Heyrovsky, Jaroslav (1890–1967) physical chemist.
Higgins, G. H.
Hildebrand, Joel Henry (1882–1983) chemist.
Hinshelwood, Sir Cyril Norman (b. 1897) chemist.

Hirsch, A.
Hisinger, Wilhelm (1766–1852) electrochemist.
Hodgkin, Dorothy Mary C. (b. 1910) chemist.
Hofmann, August Wilhelm von (1818–1892) chemist.
Hofmeister, Franz (1850–1922) physiological chemist.
Hohenheim, Theophrastus Bombast von (or Paracelsus, Philippus Aureolus) (1493–1541) physician, chemist.
Homer (800's B.C.) poet, philosopher.
Humboldt, Friedrich Heinrich Alexander, Baron von (1769–1859) naturalist, geologist.
Huizenga, J. R.
Hyatt, John Wesley (1837–1920) inventor.

Ipatieff, Vladimir Nikolaevich (1867–1952) chemist.

Jabir ibn Hayyan (Latinized to Geber) (722–815) Arabian chemist.
James, R. A.
Joliot-Curie, Irene (1897–1956) physicist.
Joliot, Jean Frederic (1900–1958) physicist.
Joule, James Prescott (1818–1889) physicist.
Julian, Percy Lavon (1899–1975) chemist.

Kalckar, Herman Moritz (b. 1908) biochemist.
Kawarski, L.
Kekule von Stradonitz, Friedrich August (1829–1896) organic chemist.
Keller, G. P., assayist.
Kelvin, Lord William Thomson (1824–1907) mathematician.
Kennedy, J. C.
Kendrew, John C., biochemist.
Kepler, Johannes (1571–1630) astronomer, mathematician.
Kerr, John (1824–1907) physicist.
Kerst, Donald William (b. 1911) physicist.
King, Edward L. (b. 1920) inorganic chemist.
Kipp, P. J. (1806–1864) apothecary.
Kirchhoff, Gustav Richard (1824–1887) physicist.
Klaproth, Martin Heinrich (1743–1817) chemist.
Klason, Johan Peter (1848–1937).
Klaus, Karl Karlovich (1796–1864) chemist.
Knipping, P., physicist.
Kolbe, A. W. Hermann (1818–1884) chemist.
Kopp, Hermann Franz Moritz (1817–1892) physical chemist.
Kornberg, Arthur (b. 1918) biochemist.
Kossel, Albrecht (1853–1927) phiosological chemist.

Kossel, Walther (b. 1888) physicist.
Kurchatov, Igor Vasilievich (1903–1960) nuclear physicist.
Kusch, Polykarp (b. 1911) physicist.

Ladenburg, Albert (1842–1911) chemist.
Laennec, Rene Theophile Hyacinthe (1781–1826) physician.
Lamb, Willis Eugene, Jr. (b. 1913) physicist.
Landau, Lev Davidovich.
Langley, Samuel Pierpont (1834–1906) physicist, astronomer.
Langmiur, Irving (1881–1957) chemist.
Laplace, Pierre Simon, Marquis de (1749–1827) mathematician, astronomer.
Lapworth, Arthur (1872–1941) chemist.
Larsh, A. E.
Latimer, W. M.
Laue, Max Theodor Felix von (1879–1960) physicist.
Lavoisier, Antoine Laurent (1743–1794) chemist.
Lawes, Sir John Bennet (1814–1900) agriculturist.
Lawrence, Ernest Orlando (1901–1958) physicist.
Le Bel, Joseph Achille (1847–1930) chemist.
Leblanc, Nicolas (1742–1806) chemist, inventor, surgeon.
Levene, Phoebus Aaron Theodor (1869–1940) biochemist.
Lewis, Gilbert Newton (1875–1946) chemist.
Libby, Willard Frank (b. 1908) physical chemist.
Liebermann, Carl (1842–1914) chemist.
Liebig, Justus, Baron von (1803–1873) chemist.
Little, Arthur Dehon (1863–1935) chemical engineer, inventor.
Livingston, Milton Stanley (b. 1905) physicist.
Lomonosov, Mikhail Vasilevich (1711–1765) chemist.
Louyet, Paulin (1818–1850) chemist.
Lowe, Thaddeus S. (1832–1913) aero pioneer, inventor.
Lowig, Carl (1803–1890) chemist.
Lullus, Raymundus (1235–1315).
Lumiere, Auguste (brother to Louis Jean).
Lumiere, Louis Jean (1864–1948) chemist.
Lynen, Feodor, chemist.

Mach, Ernst (1838–1916) physicist.
Magnus, Heinrich Gustav (1802–1870) chemist, physicist.
Malpighi, Marcello (1628–1694) physician, anatomist.
Malus, Etienne Stephen Louis (1775–1812) physicist, engineer.
Manning, Winston Marvel (b. 1909) nuclear chemist.

Mansfield, Charles Blackford (1819–1855) chemist.
Marconi, Guglielmo (1874–1937) inventor, electrical engineer.
Marinsky, Jack A.
Mark, Herman Francis (b. 1895) chemist.
Martinovics, Ignatius (1755–1795).
Martius, Carl Alexander (1838–1920) chemist.
Mayow, John (1640–1679) chemist.
McClintock, Barbara.
McCollum, Elmer Verner (b. 1879) physiological chemist.
McMillan, Edwin Mattison (b. 1907) physical chemist.
Mech, J. F.
Meitner, Lise (1878–1968) physicist.
Mendel, Lafayette Benedict (1872–1935) physical chemist.
Mendeleev, Dmitri Ivanovich (1834–1907) chemist.
Mendelssohn–Bartholdy, Paul, chemist.
Meyer, Julius Lothar (1830–1895) chemist.
Meyer, Victor (1848–1897) chemist.
Michelson, Albert Abraham (1852–1931) physicist.
Midgley, Thomas Jr. (1889–1944) chemist, inventor.
Milankovich, Milutin, physicist.
Miller, William Allen (1817–1870).
Minot, George Richards (1885–1905) physician.
Mitscherlich, Eilhardt (1794–1863) chemist.
Mohr, Karl Friedrich (1806–1879) manufacturer of pharmaceuticals.
Moissan, Ferdinand F. Henri (1852–1907) chemist.
Morgan, Leon Owen (b. 1919) chemist.
Morley, Edward Williams (1838–1923) chemist, physicist.
Moseley, Henry Gwyn–Jeffreys (1887–1915) physicist.
Mulder, Gerardus Johannes (1802–1880) chemist.
Muller, Franz Joseph, Baron von Reichenstein (1740–1825) chemist, mineralogist.
Muller, W.
Mulliken, Robert Sanderson (b. 1896) physicist, chemist.
Murphy, William Parry (b. 1892) physician.

Nef, John Ulric (1862–1915) chemist.
Nernst, Walther Hermann (1864–1941) physical chemist, physicist.
Nickles, Jerome.
Nier, Alfred Otto Carl (b. 1911) physicist.
Nilson, Lars Fredrik (1840–1899) physicist, chemist.
Nishgima, K.
Nobel, Alfred Bernhard (183–1896) chemist, engineer, inventor.

Noddack, Ida Tacke (b. 1896) chemist.
Northrop, John Howard (b. 1891) biochemist.

Ochoa, Severo (b. 1905) biochemist.
Oersted, Hans Christian (1777–1851) chemist, physicist.
Olszevski, Karol Stanislav (1846–1915) chemist.
Oparin, Alexander Ivanovich, biochemist
Oppenheimer, J. Robert (1904–1967) physicist.
Ostwalt, Wilhelm (1853–1932) chemist.

Pasteur, Louis (1822–1895) chemist, biologist.
Pauli, Wolfgang Ernst (1900–1958) physicist.
Pauling, Linus (b. 1901) chemist, biochemist
Payen, Anselme (1795–1871) chemist.
Peltier, Jean Charles Athanase (1785–1845) physicist.
Perkin, Sir William Henry (1838–1907) chemist.
Perutz, Max Ferdinand, physicist.
Petit, Alexis Therese (1791–1820) physicist.
Pettenkofer, Max Joseph von (1818–1901) bacteriologist, hygienist, physician.
Piccard, Auguste (1884–1962) physicist.
Piccard, Jean Felix (1884–1963) physicist.
Planck, Max Karl Ernest Ludwig (1858–1947) physicist.
Plato (c. 427–347 B.C.) philosopher.
Poisson, Simeon Devis (1781–1840) mathematician.
Powell, Cecil Frank (b. 1903) physicist.
Pregl, Fritz (1869–1930) chemist.
Pribil, Rudolf.
Priestly, Joseph (1733–1804) chemist.
Proust, Joseph Louis (1754–8126) chemist.
Pyle, G. L.

Rabe, Paul (1869–1952) chemist.
Raman, Chandrasekhara Venkata (b. 1885) physicist.
Ramsay, Sir William (1852–1916) chemist.
Raoult, Francois Marie (1830–1902) chemist, physicist.
Rayleigh, John William Strutt (1842–1919) mathematician, physicist, naturalist.
Reaumur, Rene Antoine Ferchault de (1683–1757) naturalist, physicist, entomologist.
Reichstein, Tadeus (b. 1897) chemist.
Reilly, Sidney (b. 1874) Russian-born British Master Spy.
Reiset, J.
Remsen, Ira (1846–1927) chemist.

Reppe, Walter (b. 1892) chemist.
Richards, Theodore William (1868–1928) chemist.
Robinson, Sir Robert (b. 1886) organic chemist.
Roeder, F. A.
Roscoe, Sir Henry Enfield (1833–1915) chemist.
Rothe, Fritz.
Royds, Thomas D. (1884–1955) physicist.
Rumford, Sir Benjamin Thompson, Count (1753–1814) physicist.
Runge, Friedlieb Ferdinand (1795–1867) chemist.
Rush, Benjamin (1745–1813) chemist.
Rutherford, Ernest, 1st Baron Rutherford of Nelson (1871–1937) physicist.

Sabatier, Paul (1854–1941) chemist.
Sakharov, Andrei (b. 1920) physicist.
Sala, Angelo (1575–1640) physician.
Sanger, Frederick (b. 1918) chemist.
Scheele, Carl Wilhelm (1742–1786) pharmacist, chemist.
Schonbein, Christian Friedrich (1799–1868) chemist.
Schotten, Carl Ludwig (1853–1910) chemist.
Schrieffer, John R.
Schrodinger, Erwin (b. 1887) physicist.
Schurer, Christoph (sixteenth century).
Schwinger, Julian Seymour, nuclear scientist.
Seaborg, Glenn T. (b. 1912) physical chemist.
Semenov, Nikolai N. (b. 1893?) chemist.
Serturner, Fredrich Wilhelm (1783–1841) chemist.
Shockley, William (1890–1953) physicist.
Siedentopf, H. F. W.
Sidgwick, Nevil Vincent (1873–1952) chemist.
Sikkeland, T.
Silber, Paul (1851–1932).
Silliman, Benjamin (1779–1864) geologist, chemist, physicist.
Smith, Edgar Fahs (1854–1928) chemist.
Smith, H. L.
Smith, John Lawrence (1818–1883) chemist.
Smithson, James (1765–1829) chemist, mineralogist.
Snow, Charles Percy (1905–1980) author, physicist.
Socrates (468/470–399 B.C.) Greek natural philosopher.
Soddy, Frederick (1877–1956) chemist.
Solvay, Ernest (1838–1922) chemist.
Soubeiran, Eugene, chemist.
Spence, R. W.

Springer, Alfred.
Stahl, Georg Ernst (1660–1734) chemist, physician.
Stanley, Wendell Meredith (b. 1904) biochemist.
Staudinger, Hermann (b. 1881) chemist.
Strassmann, Fritz (b. 1902) physicist.
Street, K. Jr.
Stromeyer, Frederich (1776–1835) chemist.
Studier, Martin Herman (b. 1917) chemist.
Sumner, William Graham
Svedberg, Theodor (b. 1884) chemist.
Szilard, Leo (b. 1898) physicist.

Tachenius, Otto (c. 1620–1690) chemist, physician.
Tamm, Igor Yevgenevich (b. 1895) nuclear physicist.
Taube, Henry (1915) inorganic chemist.
Teller, Edward (b. 1908) physicist.
Tennant, Smithson (1761–1815) chemist.
Thales of Miletus (c. 640–546 B.C.) Greek mathematician, astronomer.
Thatcher, Margaret (b. 1925) Prime Minister of England.
Thenard, Baron Louis Jacques (1777–1857) chemist.
Theophrastos of Eresos (c. 373–288 B.C.) Greek naturalist, botanist.
Thomson, Sir George Paget (b. 1892) physicist.
Thomson, Sir Joseph John (1856–1940) physicist.
Tilden, Sir William Augustus (1842–1926) chemist.
Tomonago, Sin–itiro.
Townes, Charles H., physicist.
Tswett, Mikhail Semenovich (1872–1919) botanist.
Tyndall, John (1820–1893) physicist, philosopher.

Urbain, Georges (1872–1939) chemist.
Urey, Harold Clayton (1893–1981) chemist.

Van't Hoff, Jacobus Henricus (1852–1911) chemist.
Vauquelin, Louis Nicolas (1763–1829) chemist.
Vernadsky, V. I.
Villard, P.
Voit, Karl von (1831–1908) physiologist, chemist.
Volhard, Jacob (1834–1910) chemist.
Volta, Allesandro, Count (1745–1827) physicist, inventor.
Voskresenskii, Alexander Abramovich (1809–1880).
Votocek, Emil.

Waage, Peter (1833–1900) chemist.

Waals, Johannes Diderik van der (1837–1923) physicist.
Wahl, Arthur Charles (b. 1917) nuclear chemist.
Wallach, Otto (1847–1931) chemist.
Walton, Ernest Thomas Sinton (b. 1903) physicist.
Wassermann, August von (1866–1925) physician, bacteriologist.
Watson, James Dewey (b. 1928) biologist
Weber, Wilhelm Eduard (1804–1891) physicist.
Wells, Herbert George (1866–1946) author.
Welsbach, Carl Auer von, Baron (1858–1929) chemist.
Wentorf, Robert H. Jr., physical chemist.
Werner, Alfred (1866–1919) chemist.
Wideroe, Rolf.
Wieland, Heinrich O. (1877–1957) chemist.
Wigner, Eugene Paul (b. 1902) physicist.
Wiley, Harvey Washington (1844–1930) agricultural chemist.
Wilkins, Maurice H. F (b. 1916) biophysicist.
Williamson, Alexander William (1824–1904) chemist.
Willstater, Richard (1872–1942) chemist.
Wilson, Charles Thomson Rees (1869–1959) physicist.
Windaus, Adolf (1876–1959) chemist.
Winkler, Clemens Alexander (1838–1904) chemist.
Witt, Otto Nikolaus (1853–1915) chemist.
Wohler, Friedrich (1800–1882) physician, chemist.
Wollaston, William Hyde (1766–1828) chemist, physicist.
Woodward, Robert Simpson (1849–1924) astronomer, mathematician, physicist.
Wroblevsky, Sigmund Florenty von (1845–1888) physicist.

Yagodin, Gennadi (b. 1927) chemist.
Yukawa, Hideki (b. 1907) physicist.

Zeise, William Christopher (1789–1847) chemist.
Zinin, Nikolai Nikolaevich (1812–1880) mathematician, physicist, chemist.
Zinn, Walter H. (b. 1906) physicist.
Zsigmondy, Richard Adolf (1865–1929) chemist.
Zweig, George.

Publishers Note: Should any of our readers be able to furnish information missing in this Appendix, we would greatly appreciate receiving it.

Dear Reader:

Every effort has been made to ascertain that the information contained in this booklet is correct. The chance always exists, however, that the source of the information may have been erroneous. For this reason, we would appreciate being advised of any errors you may detect. We would also welcome your suggestions for additional trivia for our second volume.

LITARVAN LITERATURE
12949 West 68th Avenue
Arvada, Colorado 80004
USA

We at LITARVAN LITERATURE trust you have enjoyed *Chemistry Trivia*. We welcome your comments and suggestions.

To order additional copies, please mail the completed order blank together with your check or money order made payable to Litarvan Literature to:

LITARVAN LITERATURE
12949 West 68th Avenue
Arvada, Colorado 80004 USA
(303) 431-4345

Please send me ____ copies of Chemistry Trivia, Book I. Enclosed is my check or money order in the amount of $____________ ($10.00 plus $2.50 each for postage and handling.)

Name:____

Address:____

City_____________________________State____________Zip__________